Algorithms for Intelligent Systems

This book series publishes research on the analysis and development of algorithms for intelligent systems with their applications to various real world problems. It covers research related to autonomous agents, multi-agent systems, behavioral modeling, reinforcement learning, game theory, mechanism design, machine learning, meta-heuristic search, optimization, planning and scheduling, artificial neural networks, evolutionary computation, swarm intelligence and other algorithms for intelligent systems.

The book series includes recent advancements, modification and applications of the artificial neural networks, evolutionary computation, swarm intelligence, artificial immune systems, fuzzy system, autonomous and multi agent systems, machine learning and other intelligent systems related areas. The material will be beneficial for the graduate students, post-graduate students as well as the researchers who want a broader view of advances in algorithms for intelligent systems. The contents will also be useful to the researchers from other fields who have no knowledge of the power of intelligent systems, e.g. the researchers in the field of bioinformatics, biochemists, mechanical and chemical engineers, economists, musicians and medical practitioners.

The series publishes monographs, edited volumes, advanced textbooks and selected proceedings.

More information about this series at http://www.springer.com/series/16171

Manjaree Pandit · Hari Mohan Dubey ·
Jagdish Chand Bansal
Editors

Nature Inspired Optimization for Electrical Power System

Springer

Editors
Manjaree Pandit
Department of Electrical Engineering
Madhav Institute of Technology and Science
Gwalior, India

Hari Mohan Dubey
Department of Electrical Engineering
Madhav Institute of Technology and Science
Gwalior, India

Jagdish Chand Bansal
South Asian University
New Delhi, India

ISSN 2524-7565 ISSN 2524-7573 (electronic)
Algorithms for Intelligent Systems
ISBN 978-981-15-4003-5 ISBN 978-981-15-4004-2 (eBook)
https://doi.org/10.1007/978-981-15-4004-2

This Springer imprint is published by the registered company Springer Nature Singapore Pte Ltd.
The registered company address is: 152 Beach Road, #21-01/04 Gateway East, Singapore 189721, Singapore

Preface

Nature-inspired optimization algorithms have become very popular during the last three decades for solving complex real-world problems. Nature-inspired optimization methods are motivated by the behaviour and problem-solving skills of living beings and have their origin in some or the other natural phenomenon. These algorithms usually mimic natural processes like mutation, selection, foraging behaviour of an organism, group movement, human intelligence or physical laws. Ease of implementation, a parallel search mechanism facilitated by a population of agents, non-dependence on nature of the objective function and ability to process non-differentiable, discontinuous and non-convex functions have resulted in the unprecedented popularity of nature-inspired techniques over other numerical methods, for solving practical problems.

The electrical power systems are highly interconnected, geographically distributed over large areas, have complex dynamic constraints and possess linear/nonlinear, static/dynamic and continuous/discrete variables. Hence, nature-inspired algorithms are particularly suitable for solving the complex power system optimization problems which are high dimensional, nonlinear, non-convex, discontinuous and have a large number of equality and inequality constraints.

This book provides a compilation of a few popular algorithms and their applications to different types of problems of the power system domain. It contains nine chapters. Chapters 1–2 deal with human intelligence-based teacher learner-based optimization, Chaps. 3–6 focus on swarm intelligence, Chap. 7 deals with differential evolution, Chap. 8 presents a real-world application of the genetic algorithm, and Chap. 9 reviews various optimization techniques.

The objective of dispatching generating units in an electrical power system is to compute an optimal generation schedule to minimize the cost without violating the operating limits. Earlier, this problem comprised mainly fossil fuel generating units. Now, the system complexity increases due to the widespread involvement of a large number of renewable distributed energy resources (DERs) which are random, uncertain and introduce discrete variables in the objective function.

Chapter 1 provides a metaheuristic technique inspired by the interaction between teacher and students in the classroom named as teacher learner-based optimization (TLBO), which is implemented for solving a complex static and dynamic economic load dispatch problem with ramp rate limits, prohibited operating zones, valve point loading as well as transmission losses. Chapter 2 presents the application of elitist TLBO to solve the problem of optimal power flow during transmission network congestion through active power rescheduling for the pool-based competitive electricity market model. Chapter 3 presents the popular particle swarm optimization (PSO) algorithm which has been implemented for determining optimal values of parameters of proportional–integral–derivative (PID) controller for the load frequency control (LFC) of single area power network. Levelized cost of energy (LCOE) concept helps in establishing the economic viability of the system with renewable energy sources over a long run and also helps in deciding a feasible tariff for the hybrid renewable energy system (HRES). In Chap. 4, PSO is used to determine LCOE for a hybrid energy system. Artificial bee colony (ABC) optimization algorithm is a well-accepted swarm intelligence-based method. It is applied for solving combined economic emission dispatch (CEED) of a hybrid thermal solar PV system in Chap. 5. Grey wolf optimization (GWO) is a novel swarm intelligence approach inspired by the hierarchical hunting mechanism of grey wolves. Chapter 6 presents the solution to a multi-objective problem for a microgrid in a dynamic environment using GWO. Microgrids are becoming popular for power supply with renewable energy resources due to their adaptability and ability to work independently. In Chap. 7, a mixed-integer differential evolution (MIDE) algorithm with continuous and binary variables is used to solve static and dynamic optimal scheduling for a microgrid. In Chap. 8, non-dominated sorting genetic algorithm-II (NSGA-II) is used to solve multi-objective reactive power management (MORPM) problem for minimization of active power losses, improvement of voltage profile and minimization of the total capacity of reactive power sources (RPS) in radial distribution systems (RDS). Short-term hydrothermal scheduling (SHTS) problem deals with the scheduling of hydro and thermal generating units to fulfil required power demand. As the operating cost of the hydro units is understandably almost negligible, the primary objective is the minimization of fuel cost of thermal units while satisfying a large number of nonlinear equality/inequality constraints associated with both hydro and thermal power generating units. Chapter 9 attempts to present a state-of-the-art review of nature-inspired algorithms applied for solving SHTS problems with different dimensions and complexities.

This book is expected to be very useful to the researchers/academicians/UG and PG students who wish to work and explore the applications of nature-inspired algorithms in the power system or any other interdisciplinary area.

We sincerely appreciate the time, effort and contribution of the authors and esteemed reviewers in maintaining the quality of the papers. Special thanks to the editors and supporting team of Springer for helping in publishing this book. We would also like to acknowledge the director and management of Madhav Institute of Technology and Science, Gwalior, India, for providing facilities and necessary support.

Gwalior, India — Manjaree Pandit
Gwalior, India — Hari Mohan Dubey
New Delhi, India — Jagdish Chand Bansal

Synopsis

Problems related to electrical power systems are usually very complex due to massive dimensions, nonlinearity, non-convexity and discontinuity associated with objective functions. Also, it has a large number of equality and inequality constraints, which give rise to a complex optimization problem which is difficult to solve using classical numerical methods. Nature-inspired optimization algorithms are found to be very effective as compared to traditional optimization methods due to their ease of implementation, population-based parallel search mechanism, non-dependence on the nature of the problem and ability to handle non-differentiable, non-convex problems. The analytical model of nature-inspired techniques mimics the natural happenings and intelligence of life forms. They are mainly based on evolution, swarm intelligence, ecology, human intelligence and physical science.

This book presents a wide range of optimization methods and their applications to different electrical power system problems such as economic load dispatch, demand supply management in microgrid, levelized pricing of energy, load frequency control and congestion management, reactive power management in radial distribution system.

Contents

1 **Teaching-Learning-Based Optimization for Static and Dynamic Load Dispatch** 1
Kavita Sharma, Hari Mohan Dubey, and Manjaree Pandit

2 **Application of Elitist Teacher–Learner-Based Optimization Algorithm for Congestion Management** 13
Rupali Parmar, Sulochana Wadhwani, and Manjaree Pandit

3 **PSO-Based Optimization of Levelized Cost of Energy for Hybrid Renewable Energy System** 31
Poonam Singh, Manjaree Pandit, and Laxmi Srivastava

4 **PSO-Based PID Controller Designing for LFC of Single Area Electrical Power Network** 43
Nikhil Paliwal, Laxmi Srivastava, and Manjaree Pandit

5 **Combined Economic Emission Dispatch of Hybrid Thermal PV System Using Artificial Bee Colony Optimization** 55
Salil Madhav Dubey, Hari Mohan Dubey, and Manjaree Pandit

6 **Dynamic Scheduling of Energy Resources in Microgrid Using Grey Wolf Optimization** 69
Salil Madhav Dubey, Hari Mohan Dubey, and Manjaree Pandit

7 **Mixed-Integer Differential Evolution Algorithm for Optimal Static/Dynamic Scheduling of a Microgrid with Mixed Generation** 83
Sunita Shukla and Manjaree Pandit

8 NSGA-II Based Reactive Power Management in Radial Distribution System Integrated with DGs 101
Himmat Singh and Laxmi Srivastava

9 Short-Term Hydrothermal Scheduling Using Bio-inspired Computing: A Review 115
Khushboo Sharma, Hari Mohan Dubey, and Manjaree Pandit

About the Editors

Prof. Manjaree Pandit received her M.Tech. degree in Electrical Engineering from Maulana Azad College of Technology, Bhopal, India, in 1989 and her Ph.D. degree from Jiwaji University Gwalior, India, in 2001. She is currently working as a Professor and Dean of Academics at the Department of Electrical Engineering, M.I. T.S., Gwalior, India. She is a senior member of the IEEE, a reviewer for several journals, and has published more than 60 papers in respected international journals. Her research interests include the integration of hybrid renewable energy sources with power grids, nature inspired algorithms, ANN and fuzzy neural network applications to electrical power systems.

Dr. Hari Mohan Dubey is an Associate Professor at Madhav Institute of Technology & Science, Gwalior, India. Dr. Dubey received his Ph.D. degree in Electrical Engineering from Rajiv Gandhi Proudyogiki Vishwavidyalaya, Bhopal, India. He is associated with various SCI journals as reviewer, and has published more than 70 research papers in various international journals/conference proceedings. His main research interests are in bioinspired algorithms and their applications to electrical engineering, particularly, power system planning and operation with the integration of renewable energy sources.

Dr. Jagdish Chand Bansal is an Associate Professor at South Asian University New Delhi and Visiting Faculty at the Department of Maths and Computer Science, Liverpool Hope University, UK. Dr. Bansal received his Ph.D. in Mathematics from the IIT Roorkee. Before joining SAU New Delhi, he worked as an Assistant Professor at ABV-Indian Institute of Information Technology and Management, Gwalior, and at BITS Pilani. He is the series editor of Algorithms for Intelligent Systems (AIS), published by Springer; the Editor-in-Chief of International Journal of Swarm Intelligence (IJSI), published by Inderscience; and an Associate Editor of IEEE ACCESS and ARRRAY (Elsevier). He is the general secretary of the Soft Computing Research Society (SCRS). His main research interests are in swarm intelligence and nature inspired optimization techniques. Recently, he proposed a fission–fusion social structure-based optimization algorithm, Spider Monkey Optimization (SMO), which is currently being applied to various problems in the engineering domain. He has published more than 60 research papers in various international journals/conference proceedings.

Chapter 1
Teaching-Learning-Based Optimization for Static and Dynamic Load Dispatch

Kavita Sharma, Hari Mohan Dubey, and Manjaree Pandit

Abstract This chapter presents a population-based meta-heuristic called teaching-learning-based optimization (TLBO) for the solution of economic scheduling of power generators. The mathematical model of TLBO basically simulates the interaction of the teacher with students in a classroom during the optimization process. To demonstrate the applicability and validity of TLBO, it has been implemented and tested on two different types of complex constrained economic dispatch problems that include test cases of distinct nature. To confirm the superiority and efficacy of TLBO over the existing approaches in terms of optimal search behavior and robustness, the outcomes of simulation results are also compared with other recent reported methods.

Keywords Economic load dispatch · Ramp rate limits · Prohibited operating zones · Teacher and learner phase

1 Introduction

Economic load dispatch (ELD) is one of the significant optimization issues of power system operation with the objective to determine the best possible output from number of available power generating unit to fulfill the specified system demand at the minimum operational cost, subjected to all associated operational limitations. It can be either static economic dispatch (SELD) or dynamic economic dispatch (DELD).

K. Sharma (✉) · H. M. Dubey · M. Pandit
Madhav Institute of Technology and Science, Gwalior, Madhya Pradesh, India
e-mail: kavita9031sharma@gmail.com

H. M. Dubey
e-mail: harimohandubeymits@gmail.com

M. Pandit
e-mail: drmanjareep@gmail.com

M. Pandit et al. (eds.), *Nature Inspired Optimization for Electrical Power System*, Algorithms for Intelligent Systems,
https://doi.org/10.1007/978-981-15-4004-2_1

The idea behind SELD is to generate a dispatch solution for a specific time period, whereas DELD refers to power dispatch by committed generators over the scheduled time period. ELD problems are quite complex to solve because of its very bigger dimension, a nonlinear objective function and variety of operational limitations. A considerable saving in cost can be achieved if the scheduling of available power generators is carried out optimally. Traditionally, the cost function is found to be pricewise quadratic while constraints are linear in nature [1]. The complexity of the problem rises significantly with the increase in power generators because of their combinatorial nature. Also in large thermal plant with a high capacity turbine, which makes the cost function non-smooth as well as non-convex in nature, here valve point effects (VPL) in problem formation provide more accurate results. But the problem with VPL effect, power balance and other operational constraints makes the problem non-differentiable non-convex and multimodal which is difficult to solve due to the presence of multiple local minima. Various efforts have been made to solve these problems till date, with constraints or multiple objectives, with the help of a variety of mathematical programming [1, 2] or by nature-inspired meta-heuristic [3–14]. Due ability to provide near global solution without any restriction of cost curve makes the nature-inspired algorithm (NIA) more popular in the area of power system operation. Well-accepted meta-heuristic includes particle swarm algorithm (PSO) [3, 4], bacterial foraging algorithm (BFA) [5], harmony search (HS) [6], biogeography-based optimization (BBO) [7], artificial bee colony (ABC) [8], Krill Herd algorithm (KHA) [9], differential evolution (DE) [10], chemical reaction optimization (CRO) [11], greedy heuristic search (GSA) [12], backtracking search algorithm (BSA) [13], CFA [14], etc. A detailed review of NIA for the solution of ELD problems can be found in Ref. [15].

Although various methods for optimization have been proposed by various researchers till date, the complication of the problem in hand needs to develop proficient algorithms for finding the best dispatch solution. In this perspective, the aim of this work is to present a new NIA for the solution of practical ELD problems, which can offer a practical option over the existing techniques.

In this chapter, a new meta-heuristic inspired by the interaction between teacher and students in the classroom called teacher–learner-based optimization is implemented for the solution of complex SELD and DELD problem related to the power system. The efficacy of the algorithm has been implemented, tested and validated on two SELD problems and a DELD problem. The distinct operational constraints include ramp rate limits, prohibited operating zones, VPL effect as well as transmission loss.

Remaining chapter is organized in the following manner: Sect. 2 presents the mathematical formulation of problem. Section 3 presents a brief description of TLBO, Sect. 4 provides simulation results, and finally, the important finding in terms of concluding remark is presented in Sect. 5.

2 Problem Statement

The DED problem is related to the minimization of objectives as total cost over the specified time frame subjected to associated operational constraints.

Cost function of the practical power generating unit is represented as a summation of both quadratic cost function and sinusoidal function.

Total cost corresponding to real power output is presented as follows:

$$F_1 = \sum_{t=1}^{T}\left\{\sum_{i=1}^{N} f_i(P_i)\right\} \quad (1)$$

$$f_i(P_i) = \left[a_i P_i^2 + b_i P_i + c_i\right] + \left|e_i \times \sin\left(f_i \times \left(P_i^{\min} - P_i\right)\right)\right| \quad (2)$$

Subjected to operational constraints:

$$\sum_{i=1}^{N} P_i - (P_{\mathcal{D}} + P_{\mathcal{L}}) = 0 \quad (3)$$

where

$$P_{\mathcal{L}} = \sum_{i=1}^{N}\sum_{\mathcal{I}=1}^{N} P_i B_{i\mathcal{I}} P_{\mathcal{I}} + \sum_{i=1}^{N} B_{i0} P_i + B_{00} \quad (4)$$

$$P_i^{\min} \le P_i \le P_i^{\max}; \forall i = 1, 2, 3 \ldots \mathrm{N} \quad (5)$$

$$-\mathcal{RRL}_{\mathrm{idown}} \le P_i^t - P_i^{t-1} \le \mathcal{RRL}_{\mathrm{iup}} \quad (6)$$

Here, P_i^{t-1} is the power output of ith unit at the prior hour, $\mathcal{RRL}_{\mathrm{iup}}$ and $\mathcal{RRL}_{\mathrm{idown}}$ are the upper and lower ramp rate limits.

$$P_i^{\min} \le P_i^t \le \underline{P}_i^1$$

$$\overline{P}_i^{\Bbbk-1} \le P_i^t \le \underline{P}_i^{\Bbbk}, \Bbbk = 2,3,\ldots,\mathbb{m}z \quad (7)$$

$$\overline{P}_i^{\mathbb{m}z} \le P_i \le P_i^{max}$$

Here, $\underline{P}_i^{\Bbbk}$ and $\bar{P}_i^{\Bbbk}$ represent the lower and upper limits of the $\Bbbk$th prohibited operation zone of the ith power generating unit.

3 Teaching–Learning-Based Optimization

Teaching–learning is a process where individual tries to learn from others to improve themselves. TLBO is a population-based meta-heuristic that simulates the interaction of teacher with students in a classroom [16]. It simulates two basic modes of the learning process. They are (a) with the help of teacher called teacher phase (b) interaction with other students called the learner phase. In this algorithm, a group of students are analogous to population size ($\mathcal{N}$), distinct subjects are analogous to design variables ($\mathcal{D}$), result of student represents fitness value of problem and as teacher is the most educated person of society, and hence, the best solution is represented by teacher. Its step-by-step process of implementation is depicted as below.

Randomly initialize the population ($\mathbb{X}$) within lower limits ($\mathbb{X}_j^{\min}$) and upper limits ($\mathbb{X}_j^{\max}$) of search space as follows:

$$\mathbb{X}_{ij} = \mathbb{X}_j^{\min} + \mathcal{R} * \left(\mathbb{X}_j^{\max} - \mathbb{X}_j^{\min}\right) \tag{8}$$

where random number $\in (0, 1)$.

Step 1: Teacher Phase

The average result ($\mathbb{M}^g$) of each subject in a class at generation g can be represented as follows:

$$\mathbb{M}^g = \left[\mathfrak{m}_1^g, \mathfrak{m}_2^g, \ldots, \mathfrak{m}_{\mathcal{D}}^g\right] \tag{9}$$

The best solution of objective function among the population is termed as teacher ($\mathbb{X}_T$) because a teacher always tries to enhance average result of each subject ($\mathbb{M}^g$) through motivating and helping individual learner ($\mathbb{X}_i$) towards ($\mathbb{X}_T$). Analytically, it can be represented as follows:

$$\mathbb{X}_{i\_\text{new}}^g = \mathbb{X}_i^g + \mathcal{R} * \left(\mathbb{X}_T^g - \mathcal{T}_f * \mathbb{X}^g\right) \tag{10}$$

where $\mathcal{T}_f$ is the teaching factor and it can be either 1 or 2. Its value is selected on a random basis with equal probability as below:

$$\mathcal{T}_f = \text{round}\{1 + \mathcal{R}(0, 1) * (2 - 1)\} \tag{11}$$

If $\mathbb{X}_{i\_\text{new}}^g$ is found to be better, it is related to $\mathbb{X}_i^g$, otherwise $\mathbb{X}_i^g$ remains unchanged.

Step 3: Learner Phase

Here, learner tries to improve their information by interaction with others. It can be selected on random basis.

$$\mathbb{X}_{i\_\text{new}}^g = \mathbb{X}_i^g + \mathcal{R} * \left(\mathbb{X}_i^g - \mathbb{X}_r^g\right) \text{ if } f(\mathbb{X}_i^g) < f\left(\mathbb{X}_r^g\right) \tag{12}$$

$$\mathbb{X}_{i\_new}^g = \mathbb{X}_i^g + \mathcal{R} * \left(\mathbb{X}_r^g - \mathbb{X}_i^g\right) if\ f(\mathbb{X}_i^g) > f\left(\mathbb{X}_r^g\right) \tag{13}$$

Its step-by-step process of implementation is depicted as below.

Step1: Initialize population randomly within upper and lower limits.
Step2: Evaluate population and arrange them on the basis of fitness value.
Step3: Modify the fitness value based on the concept of teacher phase, i.e., learning of learner by teacher.
Step4: Modify the fitness value based on the concept of learner phase, i.e., based on the concept of mutual interaction.
Step5: Repeat above steps till the termination criterion is reached.

4 Description of Problems and Simulation Results

To analyze the efficacy of the TLBO algorithm for the solution of real-world optimization, two dissimilar natures of ELD problems are taken here. Type one includes two cases of SELD of problem with six and ten generating unit systems with diverse cost curve, where optimal scheduling is carried out for specific time period, where the second case is having a DELD problem with four generators where scheduling period of thermal plants is 1 day or 24 h with 24 intervals of 1 h each.

Case 1: Six unit Problem with ramp rate limit, prohibited operating zones and losses This system has six thermal generation units. All generators have prohibited operating zones and ramp rate limits constraints [4]. Transmission loss is also considered here for simulation analysis. The power demand considered is 1263 MW. The outcomes of simulation analysis in terms of power generation schedules are tabulated in Table 1. While comparison with results with other methods for the same

Table 1 Comparison of results in terms of generation scheduling with different algorithms (Case 1)

O/P	PSO-WPF [4]	BBO [7]	MABC/D/Cat [8]	OKHA [9]	CFA [14]	TLBO
PG1	447.2547	447.3997	447.5032	447.3988	446.8623	447.0721
PG2	173.4092	173.2392	173.3177	173.2409	173.2990	173.1811
PG3	263.9369	263.3163	263.4631	263.3815	264.0771	263.9171
PG4	139.2500	138.0006	139.0650	138.9802	139.0329	139.0504
PG5	165.2494	165.4104	165.4735	165.3914	165.6988	165.5744
PG6	86.3095	87.07979	87.1355	87.0520	86.4471	86.6208
P_L (MW)	12.4100	12.446	12.9582	12.4448	12.4172	**12.4159**
Min. Cost ($/hr)	15442.6601	15443.0963	15449.8995	15,443.063	15,442.6553	**15442.6325**

test case is available in recent literature, it is clearly observed that TLBO can provide best power generation scheduling, in terms of minimum operational cost as 15442.6325 \$/hr in comparison with PSO-WPF [4], BBO [7], MABC [8], OKHA [9] and CFA [14]. The cost convergence curve obtained by TLBO algorithms is illustrated in Fig. 1 which is found to be steady and stable. The statistical analyses over the thirty repeated trails are tabulated in Table 2 in terms of generation cost and standard deviation. With comparisons of results with other methods, here it is clearly observed that even the average cost attained by TLBO is found to be superior to others.

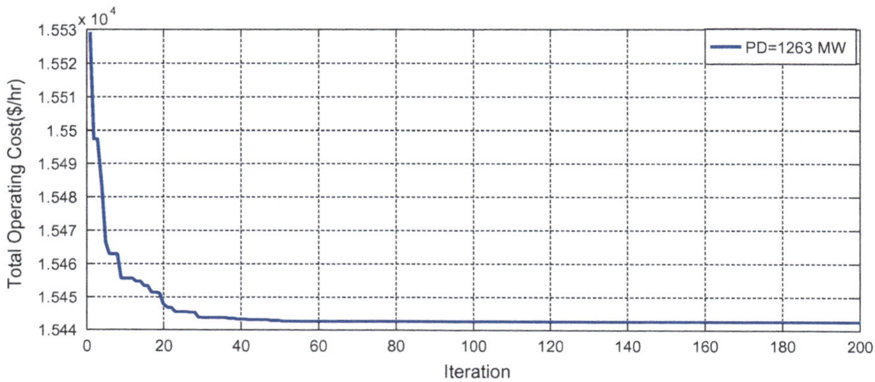

Fig. 1 Cost convergence curve obtained by TLBO for Case 1 (six unit system)

Table 2 Statistical comparison of results with other meta-heuristics (Case 1)

Methods	Generation Cost(\$/hr)			S. D	CPU time/iter (s)
	Max.	Min.	Ave.		
PSO [3]	15492	15450	15454	0.0002	14.89
GA [3]	15542	15459	15469	0.0570	41.58
PSO-WPF [4]	15442.6658	15442.6601	15442.6613	–	0.007686
ABF NM [5]	–	15 443.8164	15 446.95383	2.58223	–
HHS [6]	15453	15449	15450	–	0.14
BBO [7]	15443.096	15443.096	15443.096	–	0.0325
MABC/D/Cat [8]	15449.8995	15449.8995	15449.8995	6.04×10^{-8}	–
OKHA [9]	15,443.916	15,443.075	15,443.327	–	–
OGSA [12]	15443.06	15443.06	15443.06	000	–
CFA [14]	15,442.7002	15,442.6553	15,442.6735	0.0119	–
TLBO	15442.6701	15442.6325	15442.6552	0.0132	0.0273

Case 2: Ten unit Problem with valve point loading effects and losses This test system has ten generating units with valve point loading and transmission loss. For analysis purpose, load demands for this system are considered as 2000 MW. The fuel cost coefficient and B-loss coefficient data are adopted from ref [10]. The optimum dispatch solution obtained by simulation using TLBO algorithm is presented in Table 3, and statistical comparisons of results in terms of minimum cost, average cost, maximum cost, standard deviation (SD) of cost and average computation time with other reported methods are depicted in Table 4. While comparing results with other recently reported methods as RCCRO [11], OGHS [12], BSA [13] and ADE-MMS [10], it is clearly observed that TLBO can provide a better solution in terms of minimum cost of 111496.6215 $/hr with minimum SD 0.00910. Also, the average computation time over the 500 iterations is 4.25 s, considering the dimension and

Table 3 Comparison of results in terms of generation scheduling with different algorithms (Case 2)

O/P	RCCRO [11]	OGHS [12]	BSA [13]	ADE-MMS [10]	TLBO
PG1	55	55	55	55.00000000	55
PG2	79.9999	80	80	80	80
PG3	106.922	106.9916	106.9295	106.93993334	106.9792
PG4	100.5426	100.5354	100.6028	100.57627031	100.6464
PG5	81.5216	81.445	81.499	81.50173667	81.5942
PG6	83.0528	83.067	83.0074	83.02088436	82.8198
PG7	299.9999	299.9998	300	300	300
PG8	339.9999	339.9999	340	340	340
PG9	469.9999	470	470	470	470
PG10	469.9999	469.9999	470	470	470
P_L (MW)	87.0387	87.03890848	87.0387	87.03882468	87.0396
Min. Cost ($/hr)	111497.6319	111490	111497.6276	111497.630810	111496.62150281

Table 4 Statistical comparison of results (Case 2)

Methods	Generation cost ($/hr)			S. D	Ave CPU time (s)
	Max.	Min.	Ave.		
ADE-MMS [10]	111497.630814	111497.630810	111497.630810	0.000001	1.122402
OGHS [12]	–	111490	–	–	–
BSA [13]	111497.6307	111497.6276	111497.6286	0.0008	4.29
TLBO	111496.6474	111496.6215	111496.6257	0.00910	4.25

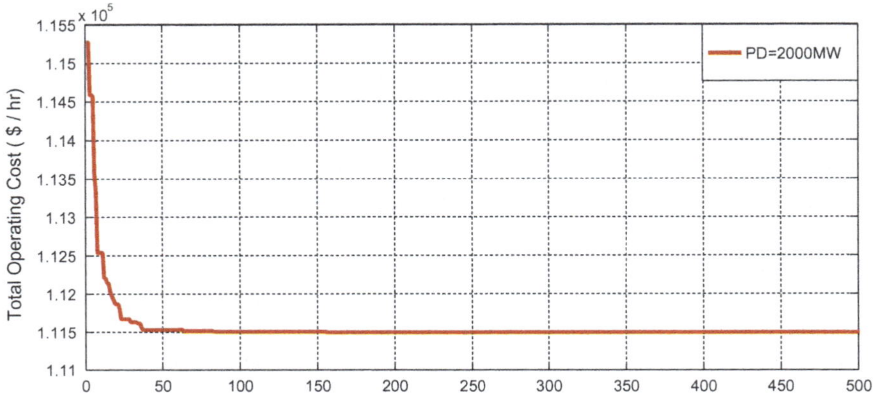

Fig. 2 Cost convergence curve obtained by TLBO for Case 2 (ten unit system)

complexity of the problem it seems to be obvious. The smooth cost convergence curve for this case obtained by TLBO is shown in Fig. 2.

Case 3: Four units Problem with convex fuel cost characteristics A four unit system is considered here for solution of DELD problem over the scheduled time period of 24 h. The fuel cost is second-order polynomial and adopted as per Ref. [17]. The load profile over the 24 h is adopted as per the same reference. The transmission loss is not considered here. The best cost obtained by TLBO is 647964.5608 $, after the thirty repeated trials which is validated by the results of GAMS [17]. The generation schedule corresponding to the best cost is tabulated in Table 5, and its statistical results are listed in Table 6. The smooth cost convergence curve of this DELD problem obtained by TLBO is presented in Fig. 3.

Effect of population

Change in population effects the performance of algorithm, i.e., high population make the algorithm sluggish and incompetent on the other hand low population may not be competent for searching a minima in case of multi model function. In most of the literature, it is clearly stated that the optimum population size is selected on the basis of problem dimension and its associated complexity. Table 7 demonstrates the performance of algorithm for different population sizes for the above mentioned three cases. After thirty careful trials, it was observed for SELD problems. Lower population size of 50 performs better and also has lower computational time, whereas for the DELD problem higher population size of 150 provides a better solution in terms of fuel cost, however computational time is high. Considering the operational constraints over the scheduled time period of 24 h, computational time of 140.6486 s seems to be quite obvious.

Solution quantity and Robustness

Considering the statistical results as tabulated in Tables 2, 4 and 6 for the abovementioned three cases, it was clearly observed that for Case 1, TLBO performs better than all reported methods and almost comparable to recently reported method as PSO-WPF [4], CFA [14].

Table 5 Generation scheduling of DELD problem (Case 3)

Hour	P_{g1}	P_{g2}	P_{g3}	P_{g4}	Pd (MW)
1	166.212	112.0672	130.4793	101.2415	510
2	172.5983	116.6751	135.4191	105.3075	530
3	168.096	113.306	132.1407	102.4573	516
4	166.339	112.0283	130.2982	101.3345	510
5	167.6596	113.1721	131.8982	102.2701	515
6	176.8916	119.9658	139.0555	108.0871	544
7	200	145.7118	169.0255	131.2627	646
8	200	159.0694	183.6179	143.3127	686
9	200	184.7486	190	166.2514	741
10	200	181.1252	190	162.8748	734
11	200	188.4965	190	169.5035	748
12	200	194.775	190	175.225	760
13	200	191.5492	190	172.4508	754
14	200	163.7382	188.8212	147.4406	700
15	200	159.154	183.5212	143.3248	686
16	200	173.7207	190	156.2793	720
17	200	170.4435	190	153.5565	714
18	200	195.24	190	175.76	761
19	200	177.3771	190	159.6229	727
20	200	168.5705	190	155.4295	714
21	198.1548	138.573	160.0014	121.2708	618
22	189.8335	128.8503	149.1501	116.1661	584
23	187.8591	127.2281	147.8707	115.0421	578
24	177.0355	119.8052	139.0598	108.0995	544
Total Cost($)					140.6486

Table 6 Statistical results for four unit DELD problem (Case3)

Methods	Generation cost ($)			S. D	Time/iteration (s)
	Max.	Min.	Ave.		
GAMS [17]	–	6.4796×10^5	–	–	–
TLBO	647972.8985	647964.5608	647968.6976	3.1137	140.6486

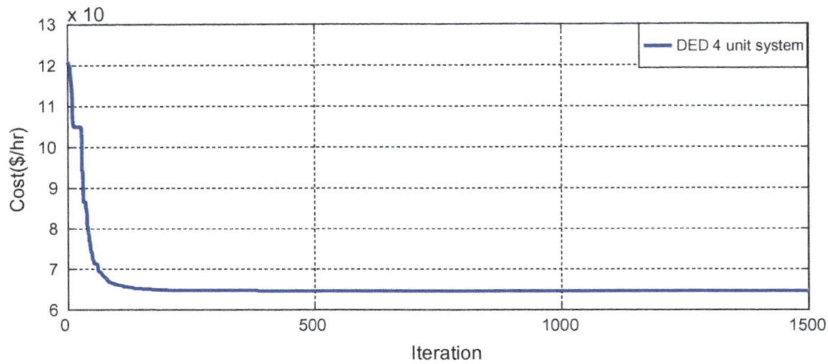

Fig. 3 Cost convergence curve obtained by TLBO for Case 3(four unit system)

Table 7 Effect of change in population size for different test cases

Test Case	NP	Max.	Min.	Ave.	SD	CPU time (s)
Case 1	50	15442.6701	15442.6325	15442.6552	0.0132	4.0531
	100	15442.6643	15442.6534	15442.65981	0.00529	10.5023
	150	15442.6639	15442.657	15442.66249	0.00157	15.5786
Case 2	50	111496.6474	111496.6215	111496.6257	0.00910	4.25
	100	111496.6313	111496.6200	111496.6226	0.00428	11.0541
	150	111496.6387	111496.6196	111496.6247	0.00529	14.1288
Case 3	50	647986.215	647964.5642	647973.1839	9.3232	55.5426
	100	647986.215	647965.5908	647972.2881	5.3599	110.7478
	150	647972.8985	647964.5608	647968.6976	3.1137	140.6486

For Case 2, generation cost attained by TLBO is found to be superior to ADE-MMS [10], OGHS [12] and BSA [13] in of terms maximum average and minimum cost.

Because of randomness is involved in generating the initial population of any heuristic search optimization techniques, single run of algorithm is not sufficient to judge the best result. Therefore, many run with different population sizes were carried out to analyze the consistency of the TLBO algorithm for all the three test cases. The statistical results in terms of cost over the thirty repeated trials have been tabulated in Table 7. Here, it was observed that the frequency of achieving optimum cost is more as the standard deviation of cost was found to be low for all the test case. Hence, we can say that TLBO is robust algorithm and has global minima searching capability too.

Computational efficiency

Tables 2, 4 and 6 present best cost achieved by TLBO for six unit system with RRL and POZ which is nonlinear and discontinuous in nature, ten unit system with

VPL effects which makes the system non-convex and multimodal and a four unit system with convex fuel cost characteristics. The minimum cost achieved by TLBO is 15442.6325 \$/hr for Case 1, 111496.6215 \$/hr for Case 2 and 647964.5608 \$ for Case 3, respectively. The performance of TLBO is almost found to be better as compared to reported results by other methods available in various recent literatures. Also, the required computational time found is to be less; therefore, we can say that TLBO is one of the computation efficient algorithms and has the global searching capability.

5 Conclusion

TLBO is a population-based meta-heuristic algorithm inspired by the influence of teacher knowledge on students in the classroom. Analytical model of optimization basically mimics the impact of teacher and marks obtained by grasping power of students. To investigate the potential of the algorithm, various realistic and nonlinear characteristic constraints such as non-smooth convex cost function, VPL effect, RRL, POZ, generator limit constraints were considered. The proposed algorithm is simple and comparatively easy to implement. The algorithm was tested on the different standard cases, and the comparison of results is made with best-known results reported by other methods in terms of cost, standard deviation and required computational time. The solutions obtained by the TLBO algorithm were found to be superior and also have smooth cost convergence characteristics. Considering all these results obtained by simulation for ELD problems with different fuel cost characteristics, dimensions, demands and coupled operational constraints, it can be concluded that TLBO performs better in an efficient manner due to unique exploitation and exploration capability.

Acknowledgements The authors acknowledge the financial support provided by AICTE-RPS project File No. 8-36/RIFD/RPS/POLICY-1/2016-17 dated 2.9.2017 and TEQIP III. The authors also thank the director and management of M.I.T.S. Gwalior, India, for providing facilities for carrying out this work.

References

1. Wood AJ, Wollenberg BF (1996) Power generation operation and control, 2nd edn. Wiley, New York
2. Lin CE, Viviani GL (1984) Hierarchical economic dispatch for piecewise quadratic cost functions. IEEE Trans Power Apparatus Syst 103:1170–1175
3. Gaing ZL (2003) Particle swarm optimization to solving the economic dispatch considering the generator constraints. IEEE Trans Power Syst 18(3):1187–1195
4. Ghorbani N, Vakili S, Sarkhosh A (2017) A new coding for solving large-scale non-convex economic dispatch problems without a penalty factor. Int J Manage Sci Eng Manage. https://doi.org/10.1080/17509653.2016.1249426

5. Panigrahi BK, Pandi VR (2008) Bacterial foraging optimisation: Nelder- Mead hybrid algorithm for economic load dispatch. IET Gener Trans Distrib 2(4):556–565
6. Fesanghary A, Ardehali MM (2009) A novel meta-heuristic optimization methodology for solving various types of economic dispatch problem. Energy 34:757–766
7. Bhattacharya A, Chattopadhyay PK (2010) Biogeography-based optimization for different economic load dispatch problems. IEEE Trans Power Syst 25(2):1064–1077
8. Secui DC (2015) A new modified artificial bee colony algorithm for the economic dispatch problem. Energy Convers Manag 89:43–62
9. Bulbul SMA, Pradhan M, Roy PK, Pal Tandra (2018) Opposition-based krill herd algorithm applied to economic load dispatch problem. Ain Shams Engineering Journal 9:423–440
10. Zhang Q, Zou D, Duan N, Shen X (2019) An adaptive differential evolutionary algorithm incorporating multiple mutation strategies for the economic load dispatch problem. Applied Soft Comput J 78:641–669
11. Bhattacharjee K, Bhattacharya A, Dey SHN (2014) Solution of economic emission load dispatch problems of power systems by real coded chemical reaction algorithm. Int J Electr Power Energy Syst 59:176–187
12. Singh M, Dhillon JS (2016) Multiobjective thermal power dispatch using opposition-based greedy heuristic search. Int J Electr Power Energy Syst 82:339–353
13. Modiri-Delshad M, Rahim NA (2016) Multi-objective backtracking search algorithm for economic emission dispatch problem. Appl Soft Comput 40:479–494
14. Ghasemi M, Ghavidel S, Aghaei J, Akbari E, Li L (2018) CFA optimizer: A new and powerful algorithm inspired by Franklin's and Coulomb's laws theory for solving the economic load dispatch problems, Int Trans Electr Energ Syst. e2536. https://doi.org/10.1002/etep.2536
15. Dubey HM, Panigrahi BK, Pandit M (2014) Bio-inspired optimization for economic load dispatch: a review. Int J Bio-Inspired Comput 6(1):07–21
16. Rao RV, Patel V (2012) An elitist teaching–learning-based optimization algorithm for solving complex constrained optimization problems. Int J Ind Eng Comput 3(4):535–560
17. Soroudi A (2017) Dynamic economic dispatch. Power Syst Optim Model GAMS. https://doi.org/10.1007/978-3-319-62350-4_4

Chapter 2
Application of Elitist Teacher–Learner-Based Optimization Algorithm for Congestion Management

Rupali Parmar, Sulochana Wadhwani, and Manjaree Pandit

Abstract Computational intelligence (CI) is a science which provides computer, the ability to learn from data or experimental observations. Nature-inspired CI techniques derive motivation from kind of natural processes for the development of algorithm. Swarm intelligent methods are based on the individual and collective behavior of swarm members which finally outcomes to global intelligent behavior. These techniques work on the principle of exploration and exploitation. Usually, algorithms involve specific parameters which are to be defined by the user. Efficiency of algorithm depends on the tuning of parameters. Improper tuning generally traps the algorithm into local optima. Teacher–learner-based optimization (TLBO) technique does not involve any such parameter. Elitism further makes the algorithm turn toward the global optima. This chapter presents a prototype based on the newly developed elitist teacher–learner-based optimization (ETLBO) technique to solve the problem of optimal power flow during network congestion by active power rescheduling. Pool-based competitive electricity market model has been used. Validity of the proposed method is tested for three different types of contingencies of modified IEEE 30 bus and IEEE 57 bus test systems. The results are compared with basic TLBO, particle swarm optimization (PSO), random search method (RSM) and simulated annealing (SA).

Keywords Elitist teacher–learner-based optimization (ETLBO) · Congestion management (CM) · Teacher phase · Learner phase · Optimal power flow · Price bids · Security limits

R. Parmar (✉) · S. Wadhwani · M. Pandit
Madhav Institute of Technology and Science, Gwalior, Madhya Pradesh, India
e-mail: rupaliparmar@gmail.com

S. Wadhwani
e-mail: dr.s.wadhwani@gmail.com

M. Pandit
e-mail: drmanjareep@gmail.com

M. Pandit et al. (eds.), *Nature Inspired Optimization for Electrical Power System*, Algorithms for Intelligent Systems,
https://doi.org/10.1007/978-981-15-4004-2_2

1 Introduction

Conservation of synchronism in the operation of power systems becomes quite challenging in competitive electricity market. All the entities of the system operate for their own benefit and profit, leading to dishonoring of operational constraints [1].

Congestion in transmission system is a state in which the scheduled power flow crosses the secured power flow limits of the network [2]. Even a small amount of congestion may lead the system to emergency state, cascading outages and even total blackout in the system. In the open-access model, the independent system operator (ISO), ensures the power flow in network to be in its operational limits and thus maintains the security and reliability of the system.

Zonal pricing, load shedding, countertrade and rescheduling of generation are some of the popular market-based methods to get rid of congestion. In market-based methods, the hike in electricity prices due to mitigation of congestion is ultimately borne by the consumers [3].

Popularly adopted congestion management (CM) schemes and mechanisms for pricing for various market models are particularly addressed in references [4–8].

Overbye et al. [9] have sketched all operational phenomena of power systems, types of contingencies and their effect on system's security and stability. Author [10] has derived AC power flow transfer distribution factor and developed a scheme to allocate real power (MW) loading of transmission lines. Differential evolution algorithm has been applied to reschedule MW generation at buses for managing congestion in reference [11]. Active power rescheduling using lion optimization algorithm has been manifested by Gope et al. [12] and by energy storage systems in [13]. Genetic algorithm (GA) is implemented for active and reactive power rescheduling for managing congestion in [14].

Kohan et al. [15] have presented a multiobjective particle swarm optimization (PSO) for active power rescheduling along with demand response program and PSO for CM is presented by Sujatha and Kamraj [16] and Mahala and Kumar [17].

Rao et al. [18] have proposed teacher–learner-based optimization technique. It is a metaheuristic population-based algorithm and derives its basic idea from the learning process in a human group. Author has validated the algorithm on different mathematical benchmark test functions as well as mechanical design problem.

Non-convex problem of economic load dispatch has been solved using TLBO with quasi-oppositional approach [19]. TLBO for CM is proposed by Verma et al. in reference [20]. Pool-based electricity market model is used for its implementation. Results are also compared with techniques implemented earlier.

Concept of elitism and its effect on TLBO is discussed in reference [21]. Proposed concept is tested on thirty-five standard mathematical benchmark test functions. With elitism, the probability of moving toward global optima increases.

In the present work, ETLBO has been proposed to remove congestion from the transmission system and thus reducing the re-dispatch cost of real power. Investigations are made on cases of contingency including (i) single line outage, (ii) increment in demand and (iii) variation in line power limits. Results are validated on modified IEEE 30 bus and modified IEEE 57 bus test systems.

2 Problem Formulation

Mitigation of congestion is effectively achieved by varying the active power output of generating units. Such variations, however, add to the total cost. This additional cost, termed as congestion cost, is to be minimized. Function statement of cost is given in Eq. (1) [16].

Congestion cost

$$\mathrm{CC_c} = \sum_{j \in N_g} \left(C_k \Delta P_{gj}^{+} + D_k \Delta P_{gj}^{-} \right) \$/\mathrm{h} \tag{1}$$

where $\mathrm{CC_c}$ is the cumulative congestion cost incurred for shift in MW power (\$/h), C_k and D_k are incremental and decremental price bids submitted by GENCOs (\$/MWh). ΔP_{gj}^{+} and ΔP_{gj}^{-} are positive and negative alteration in active power generation of units (MW), respectively.

Equality and inequality constraints to be balanced are:

2.1 Equality Constraints

$$P_{gk} - P_{dk} = \sum_{j=1}^{N_b} |V_j||V_k||Y_{kj}| \cos(\delta_k - \delta_j - \theta_{kj}); \tag{2}$$

$$Q_{gk} - Q_{dk} = \sum_{j=1}^{N_b} |V_j||V_k||Y_{kj}| \sin(\delta_k - \delta_j - \theta_{kj}); \tag{3}$$

$$P_{gk} = P_{gk}^{c} + \Delta P_{gk}^{+} - \Delta P_{gk}^{-};\ k = 1, 2, 3, \ldots N_g \tag{4}$$

$$P_{dj} = P_{dj}^{c};\ j = 1, 2, 3, \ldots N_d \tag{5}$$

Here, real and reactive power at bus k are denoted by P_{gk} and Q_{gk}, respectively; real and reactive power consumed at bus k are denoted by P_{dk} and Q_{dk}, respectively; bus voltage angles at bus j and bus k are denoted by δ_j and δ_k, respectively; admittance angle of line connected between bus k bus j is denoted by θ_{kj}; number of total buses, generator buses and load buses are denoted by N_b, N_g and $\mathrm{N_d}$, respectively; P_{gk}^{c}

and P_{dj}^{c} are initial scheduled transactions of real power generated and real power consumed at bus k [22].

2.2 Inequality Constraints

$$P_{gk}^{\min} \leq P_{gk} \leq P_{gk}^{\max}, \forall k \in N_g \tag{6}$$

$$Q_{gk}^{\min} \leq Q_{gk} \leq Q_{gk}^{\max}, \forall \mathrm{k} \in N_g \tag{7}$$

$$P_{gk} - P_{gk}^{\min} \leq \Delta P_{gk} \leq P_{gk}^{\max} - P_{gk}, \tag{8}$$

$$V_m^{\min} \leq V_m \leq V_m^{\max}, \forall m \in N_1 \tag{9}$$

$$P_{ij} \leq P_{ij}^{\max}, \tag{10}$$

Here, upper bounds and lower bounds of associated variables are denoted by superscripts min and max. P_{ij} states real power flow in line between ith and jth bus while, $P_{ij}^{\max}$ is its real power rating. N_1 defines number of load buses and V_{m} is the magnitude of voltage associated with it.

2.3 Fitness Function

Often, the objective function serves as fitness function (FF). In the present research, penalty approach [16] is followed, in which, inequality constraints behave as penalty functions and are added to the main cost function to build up complete fitness function.

Explicit fitness function for CM issue is presented as Eq. (11).

$$\mathrm{FF} = \mathrm{TC_c} + \alpha_1 \times \sum_{i=1}^{\mathrm{cong}} \left(P_{ij} - P_{ij}^{\max}\right)^2 + \alpha_2 \times \sum_{j=1}^{\mathrm{VB}} \left(\Delta V_j\right)^2 + \alpha_3 \times \left(\Delta P_g\right)^2 \tag{11}$$

where

$$\Delta V_j = \begin{cases} (V_j^{\min} - V_j); \text{ if } V_j \leq V_j^{\min} \\ (V_j - V_j^{\max}); \text{ if } V_j \geq V_j^{\max} \end{cases} \quad \Delta P_g = \begin{cases} (P_g^{\min} - P_g); \text{ if } P_g \leq P_g^{\min} \\ (P_g - P_g^{\max}); \text{ if } P_g \geq P_g^{\max} \end{cases}$$

Fitness function FF is supposed to be optimized to obtain minimum cost. cong and VB represent the total number of congested lines in the network and total number of buses with unacceptable bus voltage profile, respectively. α_1, α_2, α_3 are penalty factors. Numerical value of all three penalty factors is chosen as 10,000 [20]. Possibility of violations of inequality constraints is curbed by adding penalty terms to actual cost function.

3 Frame of Elitist Teacher–Learner-Based Optimization (ETLBO)

TLBO proposed by Rao et al. mimics the process of learning in a group. The candidate having the best performance is considered as a teacher for the rest of the group. Except this one teacher, all others are said to be learners. Teacher tries to improve the performance of learners. Also, the learners improve their performance by learning from other co-learners [18].

In this algorithm, the number of learners in the group is analogous to the population and the subjects taught are analogous to the design variables of the problem under consideration. The performance of any learner corresponds to the fitness value of the function involved. TLBO proceeds in the following two phases:

3.1 Teacher Phase

This segment of TLBO shows the influence of teacher in the learning process of group members. Precisely, the teacher tries to improve the performance of the group by enhancing the mean result of the group in the subject offered by him/her. Let the population size of the group be 'n', i.e., there are total 'n' members in the group. Also, let there be 'm' subjects taught to the group, i.e., there are 'm' design variables in the formulated problem. Now, say at ith iteration, M_{ji} is the mean result of the group in any specific subject.

Performance of all the candidates in the group considering all the subjects is evaluated. The best comprehensive performance is denoted as $X_{\text{total}-\text{kbest},\,i}$ and the candidate having this performance is denoted as kbest. As, kbest is the best solution, it is referred as teacher.

The difference between the algebraic mean performance in any specific subject and the performance of teacher in the same subject is given by

$$\text{Difference\_mean}_{j,k,i} = \text{ri}\big(X_{j,\text{kbest},i} - \text{TF}\ M_{ji}\big) \tag{12}$$

where

k = population size, $k = 1, 2, 3 \ldots n$; j = number of subjects, $k = 1, 2, 3 \ldots m$.

$X_{j,\text{ kbest},\ i}$ = best performance in subject j.
i = current iteration cycle.
TF = teaching factor, 1 or 2; ri = random number in the range [0, 1].

Important fact here is that TF is not to be given as input to the algorithm; rather the algorithm chooses its random value using the equation

$$\text{TF} = \text{round}\ [1 + \text{rand}(0, 1)\{2 - 1\}] \tag{13}$$

Using the Difference_mean$_{j,\ k,\ i}$, whole population is updated in the teacher phase according to the following equation

$$X'_{j,k,i} = X_{j,k,i} + \text{Difference}_{\text{mean}_{j,k,i}} \tag{14}$$

Each updated value is evaluated for its fitness and accepted only if it gives better performance than the previous one [18]. All the updated values of design variables are recorded and used as initial values for the next phase.

3.2 Learner Phase

In the latter segment of the algorithm, the exchange of knowledge takes place among all the learners. Any learner improves its own performance by randomly interacting with any other learner. A learner upgrades himself only if the other learner is better than him.

Let the population size be 'n', the following explanation gives an idea about how the algorithm proceeds in learner phase.

Two learners are selected randomly such as

$$X'_{\text{total}-A,i} \neq X'_{\text{total}-B,i} \tag{15}$$

Here, $X'_{\text{total-}A,\ i}$ = fitness value of solution A at the end of teacher phase.
$X'_{\text{total-}B,\ i}$ = fitness value of solution B at the end of teacher phase.
Updated value of candidate solution is evaluated as:

$$X''_{j,A,i} = X'_{j,A,i} + \text{ri}\left(X'_{j,A,i} - X'_{j,B,i}\right) \tag{16}$$

if, $X'_{\text{total-A},\ i} > X'_{\text{total-B},\ i}$ and

$$X''_{j,A,i} = X'_{j,A,i} + \text{ri}\left(X'_{j,B,i} - X'_{j,A,i}\right) \tag{17}$$

if, $X'_{\text{total-}B,\ i} > X'_{\text{total-}A,\ i}$

Here, 'ri' is random number in the range (0, 1) and $X''_{j,\ A,\ i}$ and $X''_{j,\ B,\ i}$ are updated values for candidate solutions 'A' and 'B'.

Fitness function is evaluated using the updated value and compared with the previous one. Updated value is accepted only if it gives better performance. Process is repeated till termination criterion is reached.

3.3 Elitism

Many a time, the existence of inferior solutions in the populations traps the algorithm in local optima. Rao and Patel [21] incorporated the concept of elitism in TLBO. Principle of elitism says that the worst solutions in each iterative phase are replaced by elite solutions. Elitism increases the competence of TLBO by reducing diversification and enhancing exploitation capability.

4 Elitist TLBO for Congestion Management

System database for the presented experimental work has been taken from [16] and the outcome has been compared with the outcome of [16] and [20].

4.1 About Test Systems

Proposed study and its analysis have been carried out on one medium-sized and one large-sized standard test system.

1. Medium-sized modified IEEE 30 bus system includes 06 generating units, 24 load buses and 41 transmission lines; the system has a base load of 283.4 MW and 126.2 MVAR.
2. Large-sized modified IEEE 57 bus system includes 07 generating units, 50 load buses and 80 transmission lines; the system has a base load of 1250.8 MW and 335.9 MVAR.

4.2 Line Outage Contingency: Case I

Contingency case of single line outage is simulated by the unavailability of transmission line between bus 1 and bus 2 of modified IEEE 30 bus test system. Lines 1–7 and 7–8 become overloaded by 17.251 MW and 05.979 MW, respectively. Percentage congestion is 14% and 05%, respectively, with gross congestion of 09% and 15.843 MW of real power losses in network.

Table 1 Active power flow (MW) during contingency and after CM: case I and II

CONT.	CL	Active power in line (MW)	Secured power flow in line with CM (MW)					Line limit (MW)
			ETLBO	TLBO [20]	PSO [16]	RSM [16]	SA [16]	
Case I	1–7	147.251	**129.35**	130	129.97	129.78	129.51	130
	7–8	135.979	**120.22**	120.78	120.78	120.60	120.35	130
Case II	1–2	310.9	**129.69**	130	129.97	129.91	129.78	130
	2–8	97.35	**64.17**	62.34	61.1	52.36	51.47	65
	2–9	103.5	**64.65**	65	64.67	55.43	54.04	65

CL—congested lines; *CONT*—contingency

4.3 Sudden Increment in Demand with Single Line Outage: Case II

Unexpected demand increment by 50% and unavailability of line 1–7 is simulated for this contingency. During contingency state, lines 1–2, 2–8 and 2–9 experience overloading of 181 MW, 32.35 MW and 38.5 MW, respectively, with total overloading of 251.75 MW. Network incurs 37.34 MW of real power losses.

4.4 Abrupt Line Power Limits Variation: Case III and IV

In third case study, the abrupt variation of line power limits of line 5–6 and 6–12 of modified IEEE 30 bus test system is simulated. These lines transmit power on reduced limits of 175 MW and 35 MW in place of 200 MW and 50 MW, respectively. Thus, lines get congested by 20.397 MW and 14.315 MW, respectively, with a total power violation of 9.7%. Total real power losses with congestion are 27.338 MW.

Fourth case study deals with abrupt variation in line limit of line 2–3 from 85 to 20 MW. Power flow in this line is 38.19 MW, making the line experience an overloading of 18.19 MW with power violation of almost 99%. Real power losses in contingency state are 27.338 MW.

Table 1 gives the tabular view of Sects. 4.2 to 4.4. Results of interest are bold faced.

4.5 Generation Rescheduling for CM

ETLBO algorithm is run for the problem for 120 independent trials on MATLAB 2009b platform and the best solution is suggested. 100 iteration cycles and initial population of 50 are suitable for all contingency cases. Elite size of '7' gives the best result in all cases. Mutation (random) is suggested to avoid duplicacy in the population.

4.6 ETLBO for Solution of CM Problem: Mathematical Procedure

Move 1: Read test system data, incremental and decremental bids for power generating units and their operating limits.
Move 2: Create desired contingency in the system.
Move 3: Run Newton-Raphson (N-R) power flow [23] while balancing Eqs. (2–5). Determine violation in load bus voltages and line limits.
Move 4: Randomly initiate the first set of solutions which is number of learners in the group. Each subject of each learner denotes the power to be rescheduled at each generating unit for congestion elimination.
Move 5: Run N-R power flow and evaluate fitness function using Eq. (17), for each solution set. Identify and store the elite solutions.
Move 6: Update the solution sets in 'teacher' phase and then in 'learner' phase of algorithm
Move 7: Check the feasibility of solutions and crop for limits if required
Move 8: Evaluate cost for the updated solution sets and thus determine the solution with minimum cost.
Move 9: Identify the worst solutions, replace them with elite solutions and modify for duplicacy.
Move 10: Stop, if termination criteria are reached, else re-follow algorithm from step 6.

5 Numerical Results and Analysis

Line outage (I): After rescheduling the generation, no lines in the network are found to be congested. Previously, overloaded lines 1–7 and 7–8 now have a secured power flow of 129.35 and 120.22 MW. Additional cost incurred is 483.12 $/h. These lines transmit power at 99% and 67% loading, respectively, after CM. Gross shift in generation is 23.45 MW. Total active power losses in system, post CM are 11.24 MW. Cost incurred by ETLBO is most low when observed. Tables 1, 2 and 5 and Fig. 1 show the findings pictorially.

Sudden increment in load with single line outage (II): Secured power flow in the network is achieved by removing congestion in lines 1–2, 2–8 and 8–9.

Table 2 Shift in active power generated by GENCOS for CM: case I and II

CONT	Technique	Shift in real power generated by GENCOS (MW)						Overall ΔP_g (MW)
		ΔP_{g1}	ΔP_{g2}	ΔP_{g3}	ΔP_{g4}	ΔP_{g5}	ΔP_{g6}	
Case I	**ETLBO**	**−9.52**	**12.99**	**−0.16**	**−0.022**	**0.011**	**0.35**	**23.450**
	TLBO [20]	−8.5876	12.985	0.4598	0.7289	−0.009	0.3988	23.169
	PSO [16]	−8.6123	10.4059	3.0344	0.0170	0.8547	−0.012	22.936
	RSM [16]	−8.8086	2.6473	2.9537	3.0632	2.9136	2.9522	23.339
	SA [16]	−9.0763	3.1332	3.2345	2.9681	2.9540	2.4437	23.809
Case II	**ETLBO**	**−9.520**	**76.9**	**0.06**	**42.83**	**23.06**	**16.93**	**168.38**
	TLBO [20]	−8.587	76.33	1.000	52.34	13.33	17.496	168.09
	PSO [16]	NA	NA	NA	NA	NA	NA	168.03
	RSM [16]	NA	NA	NA	NA	NA	NA	164.55
	SA [16]	NA	NA	NA	NA	NA	NA	164.53

NA—not available in referred literature; *CONT*—contingency

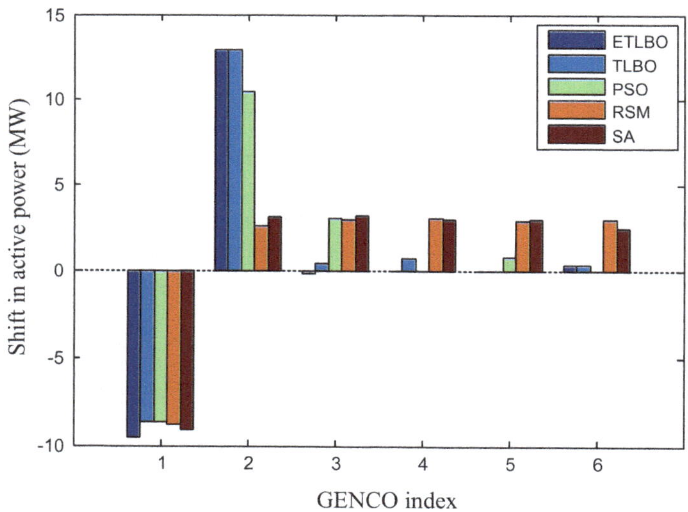

Fig. 1 Shift in active power (MW): case I

Power flow in lines is 129.7 MW, 64.17 MW and 64.65 MW, respectively. Gross shift in active power generation is 168.38 MW. Cost involved in CM by ETLBO is 5278 \$/h, which is lowest among all techniques in referred literature. These findings are also revealed in Tables 1, 2 and 5 and Fig. 2.

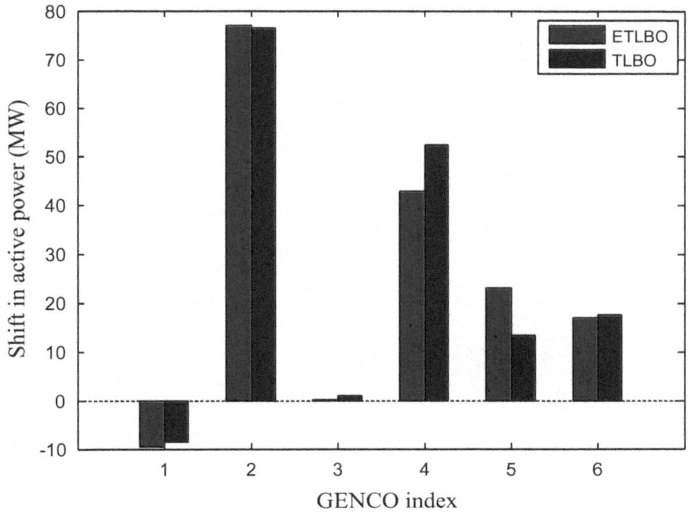

Fig. 2 Shift in active power (MW): case II

Abrupt variation in line power limits (III): The transmission network becomes free from congestion after CM. Formerly, congested lines 5–6 and 6–12 now have a secured power flow of 153.09 MW and 30.62 MW and 82% and 87% loading, respectively. System's active power losses after CM are 12.11 MW. Gross rescheduling is computed as 142.32 MW. Congestion cost is 5909.75 $/h, which is minimum among all the referred methods. Above facts can be clearly seen in Tables 3, 4, 5 and Fig. 3.

Abrupt variation in line power limits (IV): Power flow in the network has been successfully maintained within the secured operating limits of system with a CM cost of 2881.62 $/h. It is lower than all other methods in referred literature. Power

Table 3 Active power flow (MW) during contingency and after CM: case III and IV

CONT.	CL	Active power in line (MW)	Secured power flow in line with CM (MW)					Line limit (MW)
			ETLBO	TLBO [20]	PSO [16]	RSM [16]	SA [16]	
Case III	5–6	195.971	**153.09**	174.914	141	148.4	146.6	175
	6–12	49.351	**30.62**	35	34.67	35	34.84	35
Case IV	2–3	38.19	**19.43**	20	19.88	20	18.43	20

CL congested lines; *CONT* contingency

Table 4 Shift in active power generated by GENCOS for CM: case III and IV

CONT	Technique	Shift in real power generated by GENCOS (MW)							Overall ΔP_g
		ΔP_{g1}	ΔP_{g2}	ΔP_{g3}	ΔP_{g4}	ΔP_{g5}	ΔP_{g6}	ΔP_{g7}	
Case III	**ETLBO**	**34.37**	**2.71**	**0.02**	**−1.87**	**−41.69**	**−32.02**	**29.65**	**142.32**
	TLBO [20]	38.12	0.78	9.08	−0.018	−43.20	−29.90	22.81	143.92
	PSO [16]	23.14	12.45	7.49	−5.39	−81.22	0.00	39.03	168.72
	RSM [16]	59.27	0.00	37.45	−47.39	−52.12	0.00	0.00	196.23
	SA [16]	74.50	0.00	−1.52	9.95	−85.92	0.00	0.00	171.89
Case IV	**ETLBO**	**0.38**	**−27.91**	**34.63**	**0.43**	**−3.25**	**−2.87**	**−1.51**	**70.98**
	TLBO [20]	−1.02	−24.63	36.09	−6.23	−0.28	−1.26	−2.57	72.09
	PSO [16]	NA	NA	NA	NA	NA	NA	NA	76.31
	RSM [16]	NA	NA	NA	NA	NA	NA	NA	89.32
	SA [16]	NA	NA	NA	NA	NA	NA	NA	97.89

NA—not available in referred literature; *CONT*—contingency

Table 5 Comparison of cost for managing congestion for all contingencies

Study	Cost for alleviating congestion ($/h)				
	ETLBO	TLBO [20]	PSO [16]	RSM [16]	SA [16]
Case I	**483.129**	494.66	538.95	716.25	719.861
Case II	**5278.46**	5299.4	5335.5	5988.05	6068.7
Case III	**5909.75**	5981.3	6951.9	7967.1	7114.3
Case IV	**2881.32**	2916.4	3117.6	3717.9	4072.9

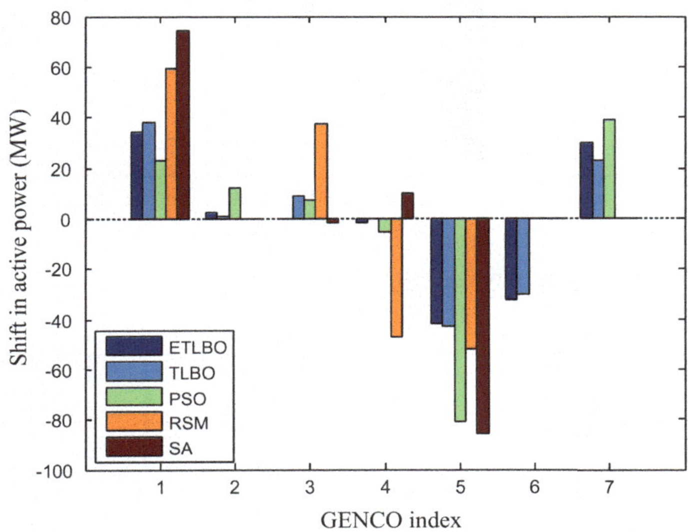

Fig. 3 Shift in active power (MW): case III

flow in congested line 2–3 is now 19.43 MW with 97% loading. System losses after CM are 20.84 MW amount of rescheduling done is 70.98 MW. This data is exhibited in Tables 3, 4, 5 and Fig. 4.

5.1 *Convergence Analysis of ETLBO*

It is studied that ETLBO performs in a stable manner and the convergence behavior is satisfactory and stable in considered cases of the proposed work. The curves obtained from experiments are shown in Figs. 5, 6, 7 and 8. All kinds of operational variations are taken into consideration.

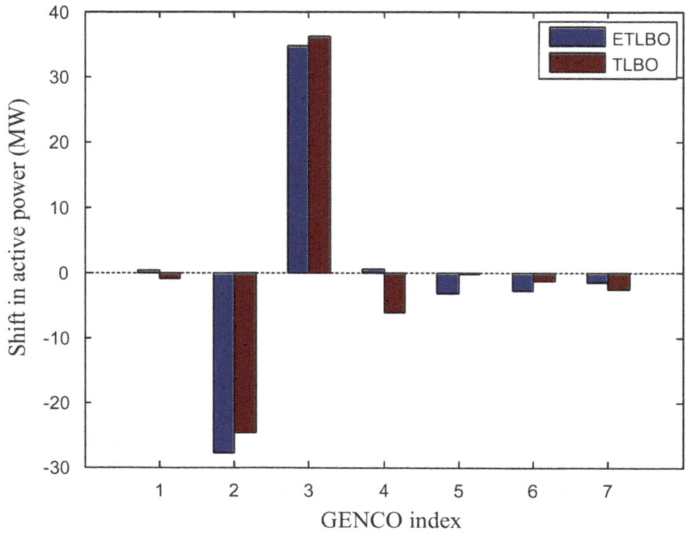

Fig. 4 Shift in active power (MW): case IV

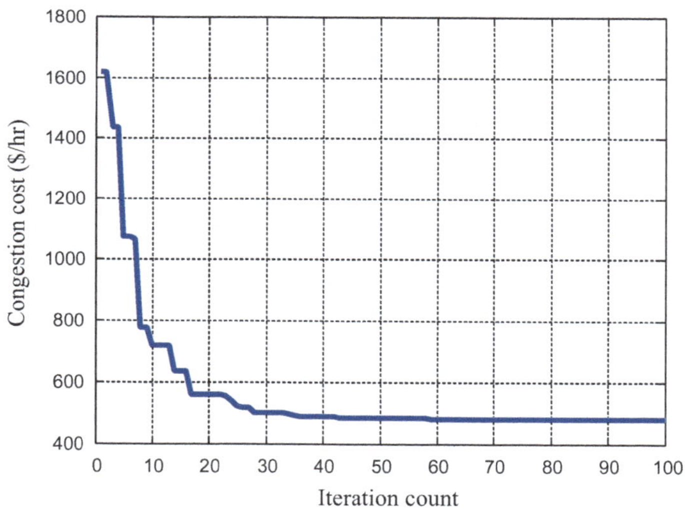

Fig. 5 Cost convergence curve: case I

6 Conclusions

The presented work confirms the competence of newly developed ETLBO over TLBO, PSO, RSM and SA for considered cases of CM during single line outage,

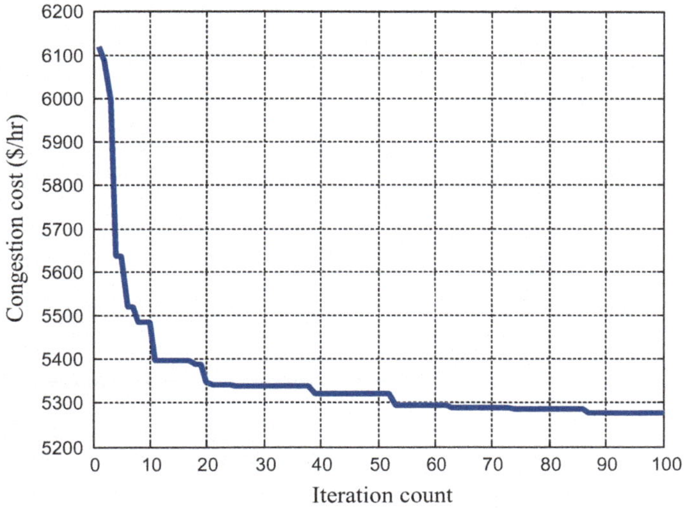

Fig. 6 Cost convergence curve: case II

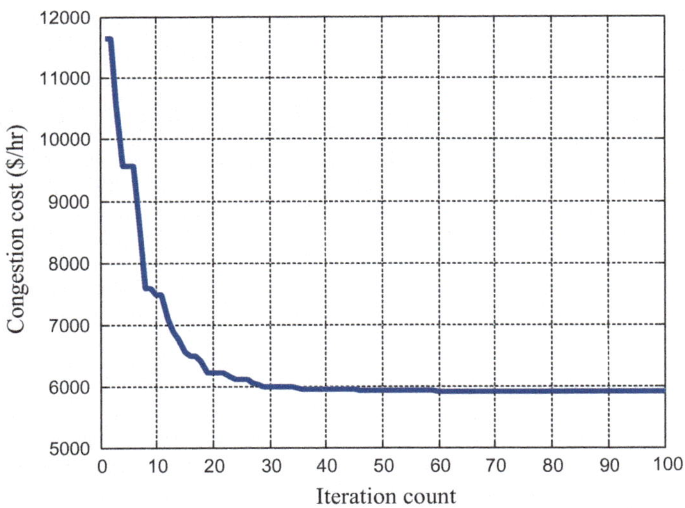

Fig. 7 Cost convergence curve: case III

load increment and abrupt variation in line flow limits. Problem of congestion in modified IEEE 30 bus system and modified IEEE 57 bus test system is used to test the performance of ETLBO technique. Pool-based competitive electricity market model is considered. Superiority of ETLBO is confirmed in terms of cost, computational

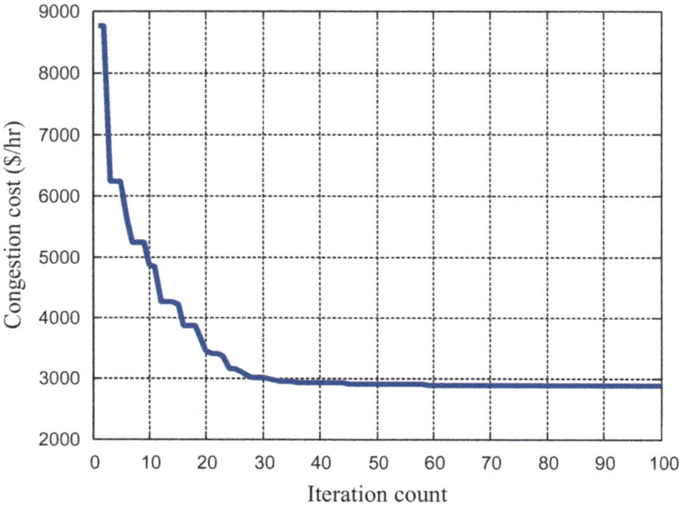

Fig. 8 Cost convergence curve: case IV

time and stability of convergence. When compared with basic TLBO, respective cost reduction in tested four cases is 11.56, 0.39, 1.2 and 1.2%.

Acknowledgements The authors acknowledge the financial support provided by AICTE New Delhi, India under the RPS research grant entitled "Addressing Power System Operational Challenges with Renewable Energy Resources Using Nature Inspired Optimization Techniques" sanctioned vide File No. 8-36/RIFD/RPS/POLICY-1/2016-17 dated August 2, 2017. The facilities and support provided by the Director and Management of M.I.T.S Gwalior, India for carrying out this work are also sincerely acknowledged.

References

1. Lai LL (2001) Power system restructuring and deregulation, 1st edn. Wiley, New York
2. Christie RD, Wollenberg BF, Wangensteen I (2000) Transmission management in the deregulated environment. P. IEEE 88(2):170–194
3. Shahidehpour M, Yamin H, Li Z (2002) Market operations in electric power systems. Wiley & Sons, Chichester
4. Bompard E, Correia P, Gross G, Amelin M (2003) Congestion-management schemes: a comparative analysis under a unified framework. IEEE Trans Power Syst 18(1):346–352
5. Faliana FD, Ilic M (1998) A mathematical framework for the analysis and management of power transactions under open access. IEEE Trans Power Syst 13(2):681–687
6. Glavitsch H, Alvarado F (1998) Management of multiple congested conditions in unbundled operation of a power system. IEEE Trans Power Syst 13(3):1013–1019
7. Pillay A, Karthikeyan SP, Kothari DP (2015) Congestion management in power systems—a review. Electrical Power Energy Syst 70:83–90

8. Kumar A, Srivastava SC, Singh SN (2005) Congestion management in competitive power market: a bibliographical survey. Electric Power Syst Res 76:153–164
9. Overbye TJ, Sauer PW, Marzinzik CM, Gross G (1995) A user-friendly simulation program for teaching power system operations. IEEE Trans Power Syst 10(4):1725
10. Kumar A, Srivastava SC, AC power transfer distribution factors for allocating power transactions in a deregulated market. IEEE Power Eng Rev, pp 42–43
11. Rajathy R, Kumar H (2012) Power flow tracing based congestion management using differential evolution in deregulated electricity market. Int J Electr Eng Inform 4(2):371–392
12. Gope S, Dawn S, Mitra R, Goswami AK, Tiwari PK (2019) Transmission congestion relief with integration of photovoltaic power using lion optimization algorithm, soft computing for problem solving. Adv Intell Syst Comput, vol 816. Springer, Singapore
13. Hemmati R, Saboori H, Jirdehi MA (2017) Stochastic planning and scheduling of energy storage systems for congestion management in electric power systems including renewable energy resources. Energy 133(C):380–387
14. Patil S, Asati N (2019) Congestion management using genetic algorithm. Int Res J Eng Appl Sci 7(2)
15. Zaeim-Kohan F, Razmi H, Doagou-Mojarrad H (2018) Multi-objective transmission congestion management considering demand response programs and generation rescheduling. Appl Soft Comput 70:169–181
16. Balaraman S, Kamaraj N (2011) Transmission congestion management using particle swarm optimization. J Electr Syst 7(1):54–70
17. Mahala H, Kumar Y (2016) Novel PSO strategy for transmission congestion management. Electr Electron Eng Int J (ELELIJ) 5(2):01–09
18. Rao RV, Savsani VJ, Vakharia DP (2011) Teaching–learning-based optimization: a novel method for constrained mechanical design optimization problems. Comput Aided Des 43:303–315
19. Prakash T, Singh VP, Singh SP, Mohanty SR (2017) Economic load dispatch problem: quasi-oppositional self-learning TLBO algorithm. Energy Syst. https://doi.org/10.1007/s12667-017-0230-3
20. Verma S, Saha S, Mukherjee V (2016) Optimal rescheduling of real power generation for congestion management using teaching learning based optimization algorithm. J Electr Syst Inform Technol. http://dx.doi.org/10.1016/j.jesit.2016.12.008
21. Rao Vivek Patel RV (2012) An elitist teaching-learning-based optimization algorithm for solving complex constrained optimization problems. Int J Ind Eng Comput 3:535–560
22. Kothari DP, Dhillon JS (2011) Power system optimization. PHI, New Delhi
23. Saadat H (2002) Power system analysis. Tata McGraw Hill Ltd., New Delhi

Chapter 3
PSO-Based Optimization of Levelized Cost of Energy for Hybrid Renewable Energy System

Poonam Singh, Manjaree Pandit, and Laxmi Srivastava

Abstract The chapter aims to optimize the levelized cost of energy (LCOE) for a sample hybrid renewable energy system (HRES) consisting of power sources such as solar photovoltaic, wind and diesel generators. The variation of life cycle cost of the system reflected by the LCOE is computed for different generation capacity factors for a time period of 24 h. The interest rate is taken as 10%, the capacity recovery factor is assumed to be 0.1175 and the life span of the hybrid generating system is considered to be 20 years. The optimal LCOE computed using a traditional solver is compared with the particle swarm optimization technique.

Keywords Hybrid renewable energy system (HRES) · Levelized cost of energy (LCOE) · Interior-point algorithm · Particle swarm optimization (PSO)

Nomenclature

CRF	Capacity recovery factor
IC	Initial capital cost (€/kW)
AE	Annual operating expenses (€/kW)
AEP	Annual energy production (kW)
i	Interest rate (%)
n	Operational life (years)
AC	The annualized costs (insurance, other expenses) (€/kW/year)

P. Singh (✉) · M. Pandit · L. Srivastava
Department of Electrical Engineering, Madhav Institute of Technology and Science, Gwalior, Madhya Pradesh, India
e-mail: lodhi.poonam18@gmail.com

M. Pandit
e-mail: manjaree_p@hotmail.com

L. Srivastava
e-mail: srivastaval@hotmail.com

M. Pandit et al. (eds.), *Nature Inspired Optimization for Electrical Power System*, Algorithms for Intelligent Systems,
https://doi.org/10.1007/978-981-15-4004-2_3

O&M	Operation and maintenance cost (€/kW/year)
CF	Net capacity factor
8760	Hours per year
$NS_1NS_2 \ldots NS_N$	Number of solar power sources for different capacity
$NW_1NW_2 \ldots NW_M$	Number of wind power sources for different capacity
$ND_1ND_2 \ldots ND_P$	Number of DG power sources for different capacity
NS_{it}	Number of solar units generating at hour '*t*'
CS_i	Capacity factor of solar *i*th unit
NW_{jt}	Number of wind power units generating at hour '*t*'
CW_j	Capacity factor of wind power *i*th unit
ND_{kt}	Number of DG units generating at hour 't'
CD_k	Capacity factor of DG *i*th unit
$PD(t)$	Demand at hour '*t*'
$PL(t)$	Losses at hour '*t*'

1 Introduction

In order to meet the rising demand of electric power along with economic considerations governed by paying capacity of consumers and environmental issues, there is a need to switch to renewable energy sources for remote area electrification in place of traditional sources. Conventional sources are depleting while renewable energy sources are non-exhaustible and can be found in abundance in our planet at particular locations. Some of the promising renewable sources of power are solar, wind, tidal and geothermal [1, 2]. The location plays a very important role in the availability of renewable sources which differs as per location. In order to meet the demand in an optimized way, the hybrid system of renewable energy sources can be used. In India, the grid-tied installed PV capacity also saw a drastic increment of around 40% between the years 2017 and 2018 from 15.7 to 22.9 GW [3–5]. This major achievement in the solar power industry is mainly due to two factors, viz. innovative technologies that are able to reduce the manufacturing costs in the past years by near about 100 times and several government schemes that are focused on providing larger incentives for the power developers and consumers [6, 7].

The LCOE is used as a measure of comparison of different electricity generating methods on a regular basis. The LCOE can also be related to the term as the average minimum price at which electricity must be sold in order to break-even over the lifetime of the project [3, 8]. The LCOE, in other words, can also be defined as the cost that can be given or assigned to every energy producing unit by the system over a predefined period, then this will be equal to the total life cycle cost (TLCC) including depreciation, maintenance cost, etc., in addition to the operating cost which may be negligible for renewable energy sources. The optimum design sizing is very important for solar-wind power generation systems with battery banks [9, 10]. The optimal sizing using an efficient optimization method can help to guarantee the lowest investment with a reasonable and full use of HRES, so that the system can work optimally with optimal configurations in terms of investment and reliability requirements for the given/forecast power demand. In this chapter, a model HRES system consisting of solar photovoltaic, wind and diesel generators have been selected for investigation. The energy output of renewable energy sources and load are dynamic in nature; hence, for meeting load demand, a conventional diesel generator is also included in this study. For the reduction of cost of energy to meet the pocket of consumers, optimization is done by reducing LCOE using traditional technique and PSO technique. Section 2 of the chapter describes the HRES and formulates LCOE; in Sect. 3, the optimization problem is formulated with equality/inequality constraints and limits; Sect. 4 discusses and summarizes the result and Sect. 5 concludes the chapter.

2 Problem Formulation

The combination of more than one renewable energy source even with the conventional source of energy called to be HRES [11, 12]. HRES is advantageous for reliability and cost to conventional source. Figure 1 shows the hybrid system of solar PV power source, wind power source and DG.

This is an assessment of the economic lifetime energy cost and lifetime energy production shown in Eqs. (1) and (2) and can be applied to essentially any energy technology [8, 13, 14].

Computational of LCOE for HRES could be written as follows:

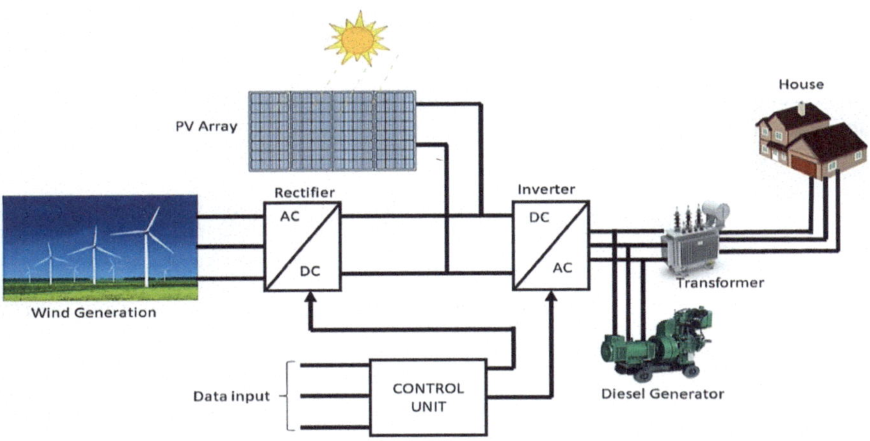

Fig. 1 PV-wind-diesel-based hybrid renewal energy system

$$\text{LCOE} = \frac{\text{Life cycle cost}}{\text{Life time energy production}}$$

$$\text{LCOE} = \frac{(\text{CRF} * \text{IC}) + \text{AE}}{\text{AEP}_{\text{net}}} \tag{1}$$

$$\text{LCOE} = \frac{\left(\left[\frac{i(1+i)^n}{(1+i)^n-1}\right] * \text{ICC} + (\text{AC} + [\text{O\&M} * n])\right)}{8760 * \text{CF}_{\text{net}}} \tag{2}$$

3 Optimization of LCOE

The objective is to minimize LCOE of HRES which consisting of solar power units, wind power units and DG units [8, 15].

$$\text{Min LCOE}(\text{NS}, \text{NW}, \text{ND.G}) \tag{3}$$

Subject to

$$\min \sum_{t=1}^{T} \text{LCOE}(\text{NS}_1\text{NS}_2 \ldots \text{NS}_N, \text{NW}_1\text{NW}_2 \ldots \text{NW}_M, \text{ND}_1\text{ND}_2 \ldots \text{ND}_P) \tag{4}$$

3.1 Power Generation Equality/Inequality Constraint

The power generated from each source must be less than or equal to the maximum capacity of the source as:

$$P_{\text{gen}} = \sum_{i=1}^{N} (\text{NS}_{it} * \text{CS}_i) + \sum_{j=1}^{M} \left(\text{NW}_{jt} * \text{CW}_j\right) + \sum_{k=1}^{P} (\text{ND}_{kt} * \text{CD}_k) - \text{PD}(t) - \text{PL}(t) \tag{5}$$

Inequality constraints

$$0 \le \text{NS}_{it} \le \text{NS}_N \quad \forall i = 1, 2 \ldots N \tag{6}$$

$$0 \leq \mathrm{NW}_{jt} \leq \mathrm{NS}_M \quad \forall j = 1, 2 \ldots M \tag{7}$$

$$0 \leq \mathrm{ND}_{kt} \leq \mathrm{NS}_P \quad \forall k = 1, 2 \ldots P \tag{8}$$

4 Results and Discussion

4.1 Test Case Description

In this chapter, LCOE is computed for a HRES with 10 solar units each of 0.53 kW, 15 wind power units each of 1.5 kW and 5 DG units each of 2.5 kW. Losses are assumed to be 5% of demand. It is assumed that the renewable units of the HRES are operating with battery support to deliver the demand at all times. However, the battery modeling is not included in this chapter for the sake of simplicity. Computation is done on an hourly basis for 24 h' time horizon. The load profile is presented in Table 1, and the cost data for the HRES is given in Table 2. Capacity recovery factor, interest rate, life span as 0.1175, 10% and 20 years for HRES, respectively.

4.2 Optimization of LCOE

Results of optimal generation allocation for the HRES with interior-point algorithm using traditional solver '*fmincon*' on MATLAB platform for one day are shown in Table 3 for capacity factors of 0.3, 0.6 and 0.8 for solar, wind power and DG units, respectively.

The results clearly indicate that the optimization is successful in minimizing LCOE with the fulfillment of equality constraint given by (5).

Table 1 Load profile for time horizon of 24 h [16]

Load (kW)	20.30	17.50	16.80	14.00	16.80	18.90	24.50	27.30
Hours	1	2	3	4	5	6	7	8
Load (kW)	25.20	22.05	21.00	21.00	21.00	21.00	20.65	21.35
Hours	9	10	11	12	13	14	15	16
Load (kW)	27.30	31.50	34.30	34.30	31.50	32.20	28.00	21.70
Hours	17	18	19	20	21	22	23	24

Table 2 Description of HRES for computation of LCOE [17]

Parameters	Cost/units
Solar PV system	
Initial investment cost (€/kW)	2832.00
Annual investment cost (€/kW)	333.00
M&O cost (€/kW/year)	56.70
Number of units	10
Wind power	
Initial investment cost (€/kW)	5832.00
Annual investment cost (€/kW)	685.02
M&O cost (€/kW/year)	116.64
Number of units	15
DG system	
Initial investment cost (€/kW)	148.00
Annual investment cost (€/kW)	70.01
M&O cost (€/kW/year)	6.4
Number of units	5

4.3 Effect of Capacity Factor on Optimal Value of LCOE

The effect of capacity factor on LCOE is analyzed by varying the capacity factor from 0.2 to 1 for solar units, wind power units and DG units. The results are plotted for the different combination cases in Figs. 2, 3, 4, 5 and 6. The LCOE is plotted for different combinations of CF values of DG and wind units while the CF of solar units is fixed at 0.2, 0.4, 0.6, 0.8 and 1.0, respectively, in Figs. 2, 3, 4, 5 and 6.

It is concluded from the results that the value of LCOE reduces as the capacity factors increase. For each case, the LCOE is least when the CF is 1.

4.4 Convergence Characteristics of the Solver

For each of the above 25 cases analyzed, 125 runs were conducted. For each case 134–135 iterations were required for convergence. Table 3 convergence behavior was obtained for each case. Figure 7 shows the convergence characteristic for $CF_s = 0.3$, $CF_w = 0.6$ and $CF_d = 0.8$. Similar curves were obtained for all other tested cases.

Table 3 Optimal power schedule of HRES for a day

Hours	PV power	Wind power	DG power	Demand	PL	Violation
1	0.207	8.859	12.249	20.300	1.015	−0.00007
2	0.203	5.865	12.307	17.500	0.875	−0.00009
3	0.135	5.402	12.103	16.800	0.840	−0.00008
4	0.083	2.239	12.378	14.000	0.700	−0.00001
5	0.132	5.378	12.130	16.800	0.840	0.00019
6	0.081	7.507	12.257	18.900	0.945	0.00018
7	0.141	13.495	12.088	24.500	1.225	−0.00017
8	1.106	15.154	12.405	27.300	1.365	−0.00011
9	0.142	14.423	11.895	25.200	1.260	−0.00008
10	0.061	10.920	12.172	22.050	1.103	−0.00010
11	0.101	9.734	12.215	21.000	1.050	−0.00044
12	0.101	9.734	12.215	21.000	1.050	0.00000
13	0.100	9.734	12.217	21.000	1.050	0.00054
14	0.102	9.733	12.215	21.000	1.050	−0.00023
15	0.111	9.338	12.233	20.650	1.032	0.00004
16	0.072	10.131	12.215	21.350	1.067	−0.00010
17	1.106	15.212	12.347	27.300	1.365	−0.00009
18	2.148	18.701	12.226	31.500	1.575	0.00004
19	1.278	22.356	12.381	34.300	1.715	−0.00017
20	1.318	22.313	12.384	34.300	1.715	−0.00020
21	2.165	18.846	12.064	31.500	1.575	0.00016
22	1.693	20.012	12.105	32.200	1.610	−0.00009
23	1.597	16.042	11.760	28.000	1.400	0.00004
24	0.058	10.558	12.169	21.700	1.085	0.00010

4.5 Validation of Results Using Particle Swarm Optimization

The results of the traditional solver are compared with PSO which is a population-based evolutionary technique. For computing the results for capacity factor 0.3, 0.6 and 0.8, the inertia constant 0.9–0.4 acceleration coefficient assumes to be 0.2 each, population size 20 and maximum no. of iteration be 100. Figure 8 shows convergence characteristics for PSO of different population size.

The evolutionary technique like PSO employ random operators, therefore, every time the algorithm is run gives slightly different results, hence, the practice is to compute result after taking few numbers of trial. Here, 10 trials are conducted and statically analyze is performed for different population size keeping the other algorithm variable fixed.

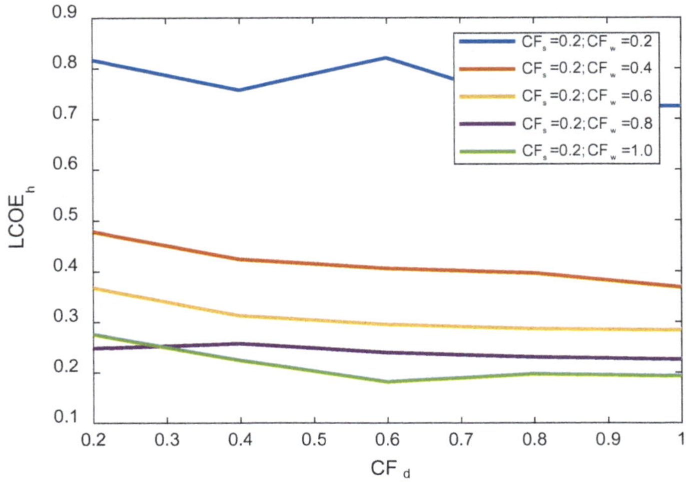

Fig. 2 LCOE of HRES with variation of CF of wind and DG ($CF_s = 0.2$)

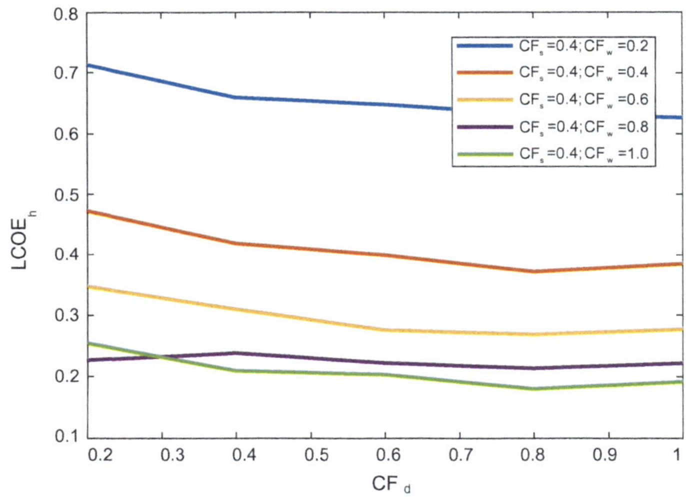

Fig. 3 LCOE of HRES with variation of CF of wind and DG ($CF_s = 0.4$)

The results are tabulated in Table 4 where it can be seen that most consistent results are obtained for population size 20 and trial 10. The consistency results for different population size are plotted in Fig. 9.

The results of traditional solver compare PSO algorithm in Table 5 and it is observed that the both results are quite close, but the traditional solver performs slightly better than PSO algorithm because PSO is random algorithm which performs better for problem with discontinuous and non-differential objective function.

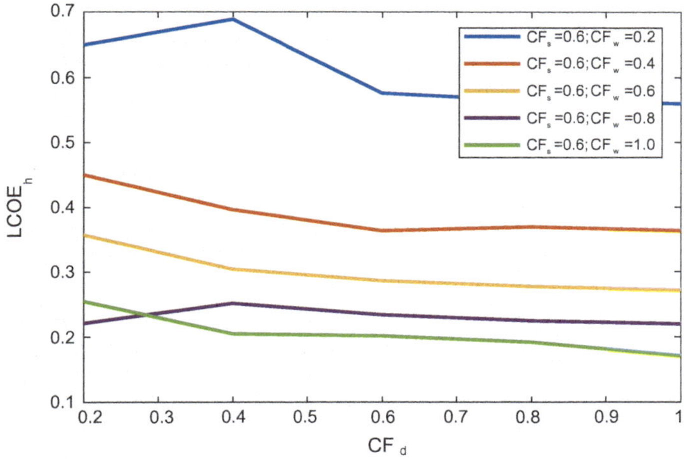

Fig. 4 LCOE of HRES with variation of CF of wind and DG ($CF_s = 0.6$)

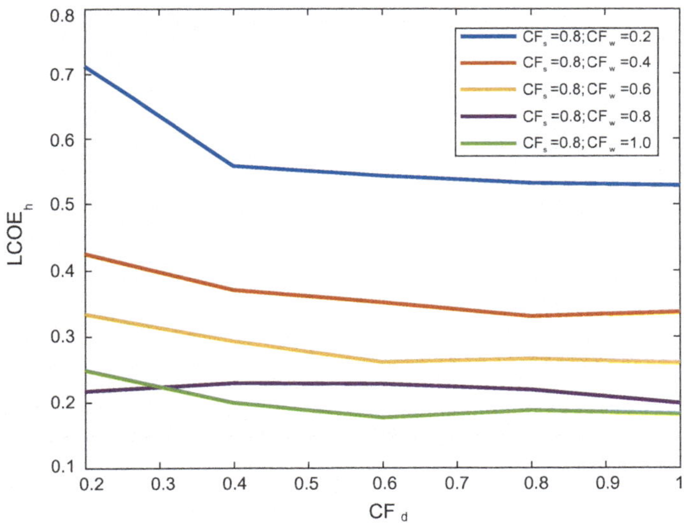

Fig. 5 LCOE of HRES with variation of CF of wind and DG ($CF_s = 0.8$)

5 Conclusion

The chapter makes an attempt to solve the optimal sizing and allocation of a hypothetical HRES with the aim to optimize the LCOE. The LCOE concept helps in establishing the economic viability of the system with renewable energy sources over a long run. This exercise also helps for deciding a feasible tariff for the HRES.

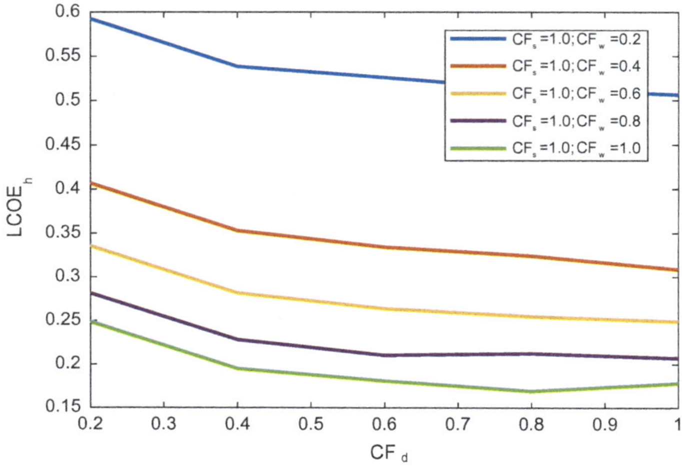

Fig. 6 LCOE of HRES with variation of CF of wind and DG ($CF_s = 1.0$)

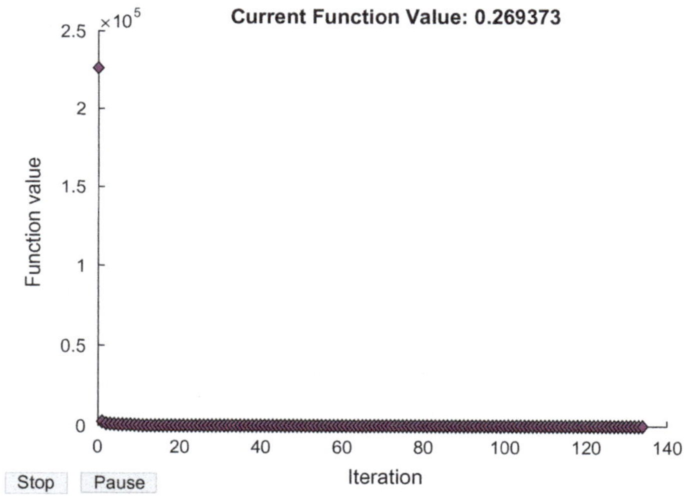

Fig. 7 Convergence characteristics

The optimal LCOE is computed using a traditional solver and the results are compared and validated using an evolutionary algorithm. The effect of capacity factor of the various generating units on the life cost is analyzed. The effect of population size on the performance is studied for the along with the convergence property. The results of both algorithms are found to be quite close. The study is expected to be useful for the emerging HRES worldwide.

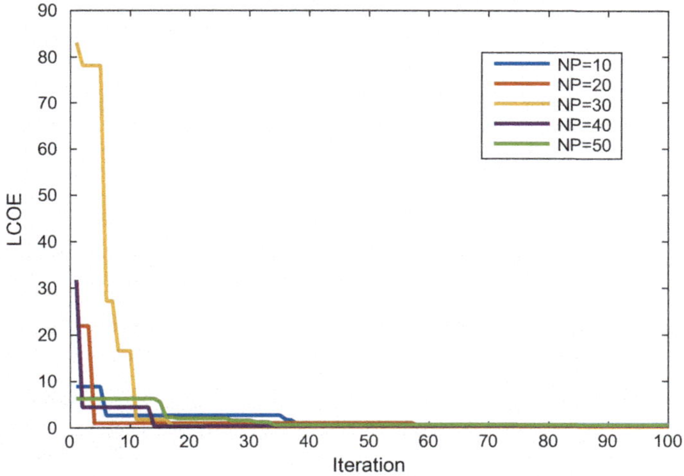

Fig. 8 Convergence characteristic for different population sizes

Table 4 Optimal data LCOE for listed population size and trial = 10

Pop size	Max value	Min value	Mean value	SD
10	0.6673	0.2244	0.4789	0.0137
20	0.5822	0.2244	0.3820	0.0115
30	0.6516	0.2370	0.4479	0.0142
40	0.6743	0.2278	0.4325	0.0120
50	0.6743	0.2621	0.4569	0.0100

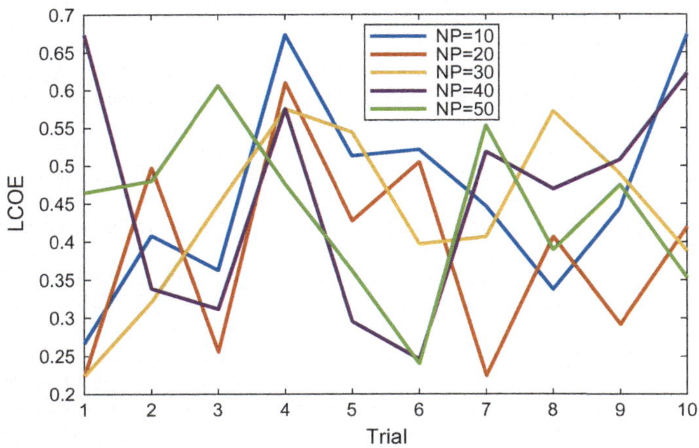

Fig. 9 Consistency comparison of PSO for different population sizes (NP)

Table 5 Comparison of results of traditional solver with PSO

Technique	LCOE (€/kW)
PSO	0.2500
Traditional solver	0.2489

References

1. Tran TTD, Smith AD (2018) Incorporating performance-based global sensitivity and uncertainty analysis into LCOE calculations for emerging renewable energy technologies. J Appl Energy 216:157–171
2. Sunderland K, Narayana M, Putrus G, Conlon M (2016) Levelised cost of energy analysis: acomparison of urban (micro) wind turbines and solar PV systems. In: 51st International universities power engineering conference (UPEC), pp 1–6
3. Baum S, von Kalben C, Maas A, Stadler I (2018) Analysis and modelling of the future electricity price development by taking the levelized cost of electricity and large battery storages into account. In: 7th International energy and sustainability conference (IESC), pp 1–8
4. Gupta RA, Kumar R, Bansal AK (2015) BBO-based small autonomous hybrid power system optimization incorporating wind speed and solar radiation forecasting. Renew Sustain Energy Rev: 41:1366–1375.
5. Fathima AH, Palanisamy K (2015) Optimization in microgrids with hybrid energy systems—a review. Renew Sustain Energy Rev 45:431–446
6. Yang H, Lu L, Zhou W (2007) A novel optimization sizing model for hybrid solar-wind power generation system. J Solar Energy 81(1):76–84
7. Javed MS, Song A, Ma T (2019) Techno-economic assessment of a stand-alone hybrid solar-wind-battery system for a remote island using genetic algorithm. J Energy 176:704–717
8. Amer M, Namaane A, M'Sirdi NK (2013) Optimization of hybrid renewable energy systems (HRES) using PSO for cost reduction. Energy Procedia 42:318–327
9. Bilal BO, Sambou V, Kébé CF, Ndiaye PA, Ndongo M (2012) Methodology to size an optimal stand-alone PV/wind/diesel/battery system minimizing the levelized cost of energy and the CO_2 emissions. Energy Procedia 14:1636–1647
10. Meena NK, Kumar A, Singh AR, Swarnkar A, Gupta N, Niazi K, Kumar P, Bansal R (2019) Optimal planning of hybrid energy conversion systems for annual energy cost minimization in Indian residential buildings. Energy Procedia 158:2979–2985
11. Ghorbani N, Kasaeian A, Toopshekan A, Bahrami L, Maghami A (2018) Optimizing a hybrid wind-PV-battery system using GA-PSO and MOPSO for reducing cost and increasing reliability. Energy 154:581–591
12. Guangqian D, Bekhrad K, Azarikhah P, Maleki A (2018) A hybrid algorithm based optimization on modeling of grid independent biodiesel-based hybrid solar/wind systems. Renew Energy 122:551–560
13. Kennedy J, Eberhart R (1996) Particle swarm optimization. In: Proceedings of IEEE International conference on neural networks 4:1942–1948
14. Ma T, Javed MS (2019) Integrated sizing of hybrid PV-wind-battery system for remote island considering the saturation of each renewable energy resource. J Energy Convers Manage 182:178–90
15. Irena (2012) Working chapter, renewable energy technologies: cost analysis series. International Renew Energy Agency, Volume 1: power sector, solar photovoltaics, June 2012
16. Borowy BS, Salameh ZM (1996) Methodology for optimally sizing the combination of a battery bank and PV array in a wind/PV hybrid system. IEEE Trans Energy Convers 11(2):367–375
17. Huneke F, Henkel J, Benavides GJA, Erdmann G (2012) Optimisation of hybrid off-grid energy systems by linear programming. Energy, Sustain Soc 2(1)

Chapter 4
PSO-Based PID Controller Designing for LFC of Single Area Electrical Power Network

Nikhil Paliwal, Laxmi Srivastava, and Manjaree Pandit

Abstract In this chapter, particle swarm optimization (PSO) technique has been implemented for determining optimal values of parameters of proportional–integral–derivative (PID) controller for the load frequency control (LFC) of single area power network. In this chapter, in place of considering one or two criteria, all the four performance indices—integral of time weighted squared error (ITSE), integral of time weighted absolute error (ITAE), integral of square error (ISE) and integral of absolute error (IAE)—have been considered as the objective functions for solving LFC problem. The implemented technique has been validated with genetic algorithm (GA) technique to show the effectiveness and applicability in tuning the PID controller. The results show the effectiveness and applicability of the proposed technique in terms of performance indices, undershoot, overshoot and settling time.

Keywords LFC · GA · PSO · Single area power network · PID controller · ITAE · ISE · IAE · ITSE

1 Introduction

Power system is dynamic, complex and large electrical network having many generators of different types, transformers, transmissions line and other electrical components. In any power system, generators are located at suitable locations, while loads are distributed through the network [1]. All the electrical loads including commercial

N. Paliwal (✉) · L. Srivastava · M. Pandit
Department of Electrical Engineering, Madhav Institute of Technology and Science, Gwalior, Madhya Pradesh, India
e-mail: pnikhil02@gmail.com

L. Srivastava
e-mail: laxmigwl@gmail.com

M. Pandit
e-mail: manjaree_p@mitsgwalior.in

M. Pandit et al. (eds.), *Nature Inspired Optimization for Electrical Power System*, Algorithms for Intelligent Systems,
https://doi.org/10.1007/978-981-15-4004-2_4

and industrial perform satisfactorily when they operate at rated frequency and voltage. But, in modern power system, there is deviation in frequency whenever loading patterns change. This deviation in frequency affects the operation and efficiency of power system along with power quality [2]. Hence, to control frequency in the power network, load frequency control (LFC) loop is used. LFC loop is designed to automatically control or adjust the frequency of the power network. The main role of load frequency controller is to sustain the frequency of power system constant or within the specified limits.

Whenever there is any unbalancing between demand and electric power generation, deviation in frequency takes place [3]. The generating unit tries to bring back the frequency to the preset or scheduled value after disturbance with the help of real power regulation [4].

For regulation of frequency in a specific area, LFC is responsible [5]. In recent times, due to the increasing numbers of interconnected power networks, changing of structure and occurrence of different types of disturbances, the LFC is becoming more and more significant [6]. The researchers are proposing miscellaneous approaches for LFC of power network to maintain frequency of system at their specified values under normal operating condition as well as during the load perturbations [7].

In [8], an elucidative review of literature on the LFC of power network has been presented. It is the fact of observing that notably work is done by the researchers for better LFC systems which is based on the neural network [9], fuzzy system theory [10], modern control theory [11] and many more. These types of advanced methods are however difficult to understand and requires familiarity of users to use mentioned techniques and hence lessen their accountability and applicability. As other option, a proportional–integral–derivative (PID) controller remains researcher's favourable choice due to its feasibility, easy implementation and also due to the appreciative ratio between cost and performances. In addition to the above-mentioned points, it also offers lower user skill requirements, simplified dynamic modelling and very little development effort [7].

Several optimization techniques like differential evolution (DE), evolution programming (EP), genetic algorithm (GA), evolution strategy (ES), particle swarm optimization (PSO), artificial bee colony (ABC) algorithm, bacterial foraging optimization (BFO) and simulated annealing (SA) have emerged in the past two decades. Due to good prospective for global optimization and several other features, GA has received importance in control system. However, GA has one of the main drawbacks of premature convergence due to which its search capability and performance reduced. As a solution to the above-mentioned drawback and problem, PSO is proposed. PSO was first introduced by Eberhart and Kennedy and it is one of the modern heuristics algorithm.

In this chapter, PSO technique has been implemented for optimal designing of PID controller for the LFC of single area power network, considering all the four performance indices, namely integral of time weighted squared error (ITSE), integral of time weighted absolute error (ITAE), integral of square error (ISE) and integral of absolute error (IAE) as objective functions, rather than considering only one or two criteria.

2 Problem Formulation

2.1 System Description

In large numbers, the various types of load are associated with the power network. Due to the load change, turbine speed and hence frequency of power network is altered. The required attribute of power network is to maintain frequency as constant as possible.

A single area power network incorporates specifically a turbine, a governor and the load with feedback speed regulator [12]. Block diagram of a single area power network with implementation of PID controller is shown in Fig. 1.

The transfer function characterization of the blocks of a single area power network is mentioned below in Eqs. 1–3, [13–15].

- Governor with dynamics:

$$G_{\mathrm{G}}(s) = \frac{1}{1 + sT_{\mathrm{G}}} \tag{1}$$

- Turbine with dynamics:

$$G_{\mathrm{T}}(s) = \frac{1}{1 + sT_{\mathrm{T}}} \tag{2}$$

- Load and machines dynamics:

$$G_{\mathrm{L}}(s) = \frac{1}{1 + sT_{\mathrm{P}}} \tag{3}$$

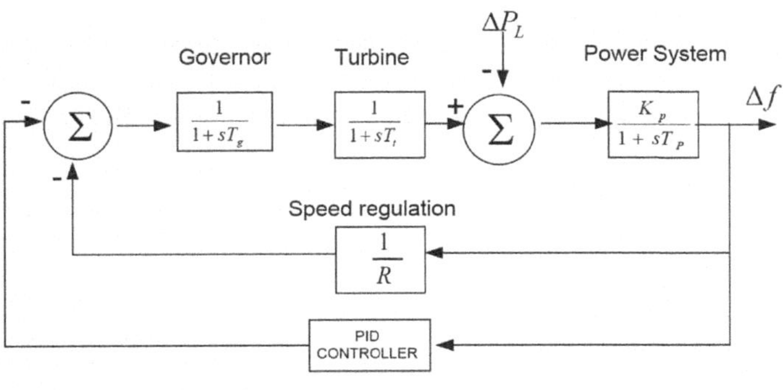

Fig. 1 Single area power network with PID controller

2.2 A Brief Introduction of PID Controller

A PID is a feedback control loop mechanism. The PID controller transfer function representation is mentioned in Eq. 4.

$$G_c(s) = K_{\mathrm{P}} + \frac{K_{\mathrm{I}}}{s} + K_{\mathrm{D}}s \tag{4}$$

where K_{I}, K_{P} and K_{D} are the coefficients for the integral, proportional and derivative terms, respectively.

2.3 Objective Function Formulation

In [15], the performance index which is considered as the objective function for LFC of single area power network is ISE. In this chapter, in place of considering one or two criteria, all the four performance indices, namely IAE, ISE, ITAE and ITSE, are considered as the objective functions.

In this chapter, PSO techniques are implemented to minimize ISE, IAE, ITSE as well as ITAE performance indices. To validate the results, GA is also implemented to minimize ISE, IAE, ITSE as well as ITAE performance indices. The four performance indices can be mathematically formulated as:

(i) $\mathrm{ITSE} = \int_0^{T_{\mathrm{sim}}} (\Delta f^2) \cdot t\mathrm{d}t$
(ii) $\mathrm{ITAE} = \int_0^{T_{\mathrm{sim}}} (\Delta f) \cdot t\mathrm{d}t$
(iii) $\mathrm{IAE} = \int_0^{T_{\mathrm{sim}}} (\Delta f)\mathrm{d}t$
(iv) $\mathrm{ISE} = \int_0^{T_{\mathrm{sim}}} (\Delta f^2)\mathrm{d}t$

where frequency deviation in the network is denoted by Δf and time range of simulation is denoted by T_{sim}.

In LFC problem, the limits of PID controller parameters are the constraints of problem [16]. Therefore, considering problem constraints, the design problem for load frequency controller can be designed as optimization problem and expressed as:

Minimize J
Subject to

(i) $K_{\mathrm{Pmax}} \geq K_{\mathrm{P}} \geq K_{\mathrm{Pmin}}$
(ii) $K_{\mathrm{Imax}} \geq K_{\mathrm{I}} \geq K_{\mathrm{Imin}}$
(iii) $K_{\mathrm{Dmax}} \geq K_{\mathrm{D}} \geq K_{\mathrm{Dmin}}$

where J implies objective function and $K_{P\mathrm{max}}$ and $K_{P\mathrm{min}}$, $K_{I\mathrm{max}}$ and $K_{I\mathrm{min}}$, $K_{D\mathrm{max}}$ and $K_{D\mathrm{min}}$ refer to the maximum and minimum values of PID controller parameters.

3 Employed Optimization Techniques

In this chapter, two optimization techniques, namely PSO and GA, are employed for LFC problem of single area electric power network.

3.1 GA

It is basically a probabilistic search heuristic algorithm very similar to natural selection mechanics and the survival of fittest [7].

3.2 PSO

It is a stochastic optimization technique which is based on the population and inspired by fish schooling's social behaviour as well as bird flocking's social behaviour. PSO is very much less susceptible to getting deceived on local optima unlike SA, GA, etc. PSO is developed in multidimensional space through bird flocking simulation [17]. The flowchart of the PSO algorithm is as shown in Fig. 2.

The PSO algorithm search for optimum value is created by the possible solutions of the problem using a swarm or group. The possible solutions are called particles. Each and every present position of particles is perceived by previously acquired position and the information of the present velocity.

The velocities of particles in PSO algorithm are updated according to the Eq. 5.

$$V_i^{k+1} = CF \times \left[V_i^k + C_1 \times rand_1 \times \left(pbest_i - s_i^k\right) + C_2 \times rand_2 \times \left(gbest - s_i^k\right)\right. \tag{5}$$

where $CF = \frac{2}{\left|2\emptyset - \sqrt{\emptyset^2 - 4\emptyset}\right|}, \emptyset = C_1 + C_2, \emptyset > 4.$

4 Results and Discussions

In this chapter, in place of considering one or two criteria, all the four performance indices, namely IAE, ISE, ITAE and ITSE, have been considered as the objective functions for solving LFC problem of single area power network. A single area thermal power plant is considered to implement PSO and GA techniques for LFC problem. In MATLAB and Simulink environment, the system is developed which is as shown in Fig. 3. The optimization algorithms, GA and PSO, are written distinctly in *.m file*.

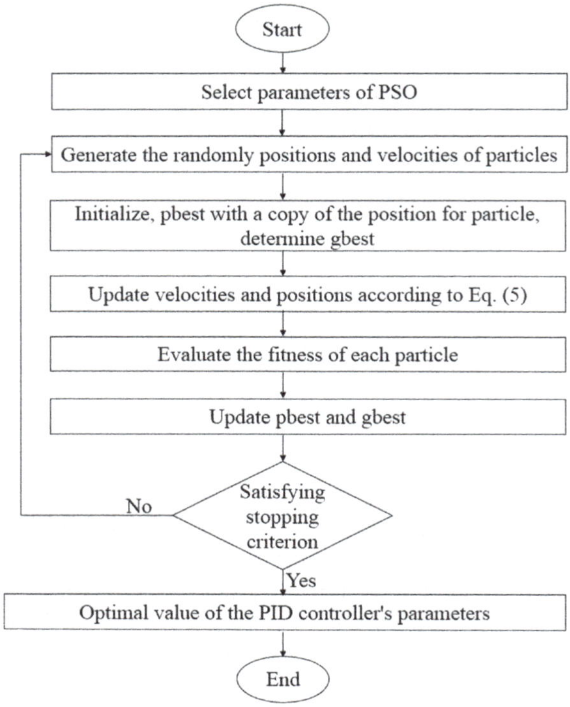

Fig. 2 Flowchart of PSO Algorithm

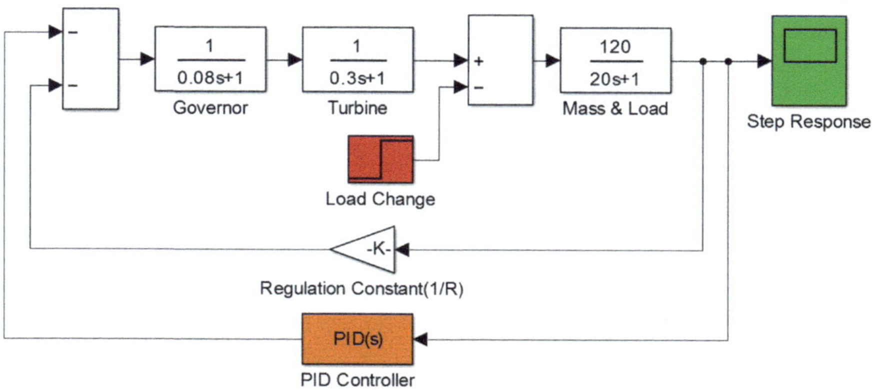

Fig. 3 Simulink block diagram of LFC with PID

The swarm size, self-adjustment, social adjustment and CF for PSO algorithm are chosen as 10, 1.49, 1.49 and 0.73, respectively [17]. The maximum number of iteration considered is 50.

The crossover fraction and stall generation limit of GA are considered as 0.650 and 125, respectively. The maximum number of iteration considered is 50.

The values of system parameters are taken from [14, 17, 18].

In this chapter, the performance analysis between an optimal PID controller based on PSO and GA techniques is done. The performance analysis here is done in terms of value of performance indices, settling time, undershoot and overshoot for 10% change in load.

The result section is divided into four different cases. In the first case, performance index IAE is objective function. In second case, performance index ISE is objective function. In third and fourth case, ITSE and ITAE are considered as objective function, respectively.

4.1 Case 1: Objective Function—IAE

In this case, performance index IAE is objective function which is subjected to minimization. The step response of the system for 10% load change using GA-PID controller and PSO-PID is as shown in Figs. 4 and 5, respectively.

The performance index, overshoot, undershoot and settling time using GA-PID are 0.050025, 0, −0.083766 and 5 s, respectively, for 10% change in load in the system.

The performance index, overshoot, undershoot and settling time using PSO-PID are 0.04999, 0, −0.081529 and 5 s, respectively, for 10% change in load in system.

The optimal value of K_P, K_I and K_D, corresponding values of IAE, overshoot, undershoot and settling time using GA-PID and PSO-PID are shown in Table 1 for 10% change in load, respectively.

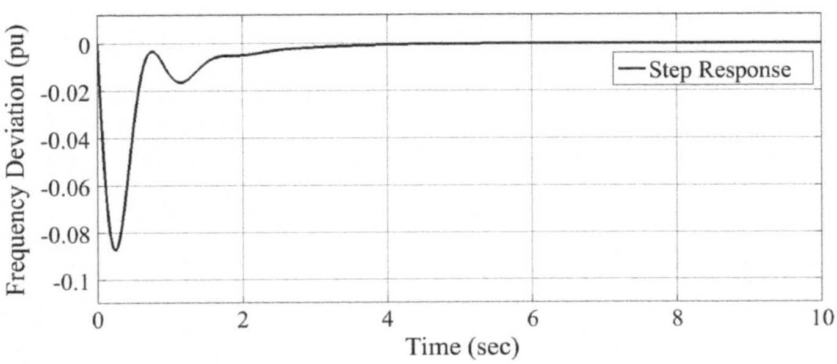

Fig. 4 GA-PID-based step response for 10% load change (case 1)

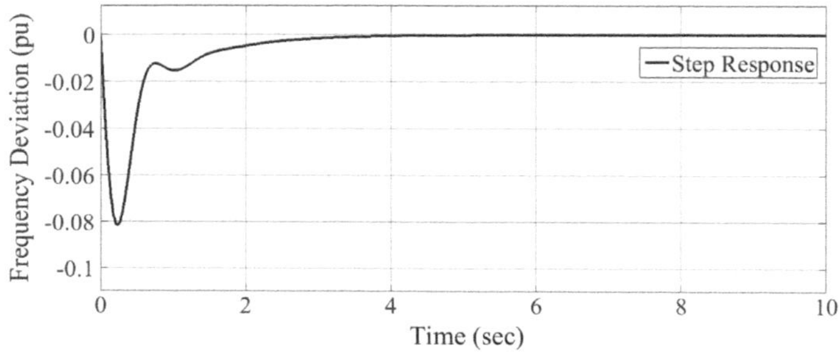

Fig. 5 PSO-PID-based step response for 10% load change (case 1)

Table 1 Comparison of GA-PID and PSO-PID for 10% load change (case 1)

	K_P	K_I	K_D	IAE	Overshoot	Undershoot	Settling Time
GA-PID	1.9912	1.9990	0.3484	0.050025	0	-0.087366	5 Sec
PSO-PID	2.0000	2.0000	0.2865	0.04999	0	-0.081529	5 Sec

4.2 *Case 2: Objective Function—ISE*

In this case, performance index ISE is objective function which is subjected to minimization. The step response of the system for 10% load change using GA-PID controller and PSO-PID is shown in Figs. 6 and 7, respectively.

The performance index, overshoot, undershoot and settling time using GA-PID are 0.001318, 0.003342, −0.035966 and 12 s, respectively, for 10% change in load in the system.

The performance index, overshoot, undershoot and settling time using PSO-PID are 0.001272, 0.003602, −0.034932 and 12 s, respectively, for 10% change in load in the system.

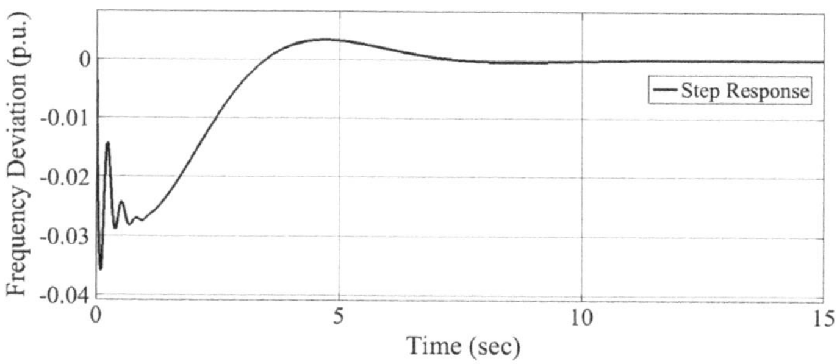

Fig. 6 GA-PID-based step response for 10% load change (case 2)

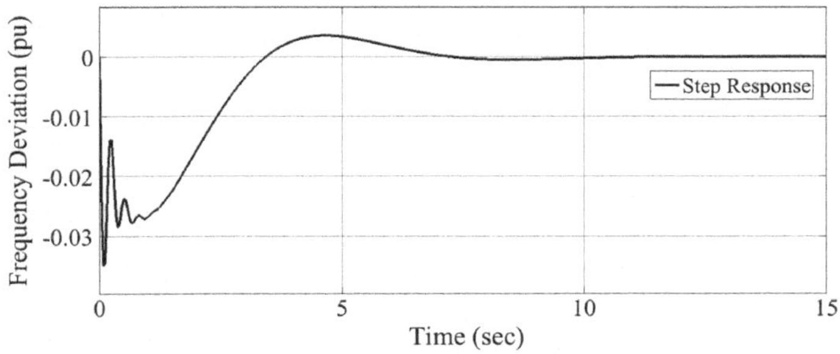

Fig. 7 PSO-PID-based step response for 10% load change (case 2)

Table 2 Comparison of GA-PID and PSO-PID for 10% load change (case 2)

	K_P	K_I	K_D	ISE	Overshoot	Undershoot	Settling Time
GA-PID	1.9970	1.9315	1.9374	0.001318	0.003342	-0.035966	12 Sec
PSO-PID	2.0000	2.0000	1.9990	0.001272	0.003602	-0.034932	12 Sec

The optimal value of K_P, K_I and K_D, corresponding values of ISE, overshoot, undershoot and settling time using GA-PID and PSO-PID are shown in Table 2 for 10% change in load, respectively.

4.3 *Case 3: Objective Function-ITAE*

In this case, performance index ITAE is objective function which is subjected to minimization. The step response of the system for 10% load change using GA-PID controller and PSO-PID is shown in Figs. 8 and 9, respectively.

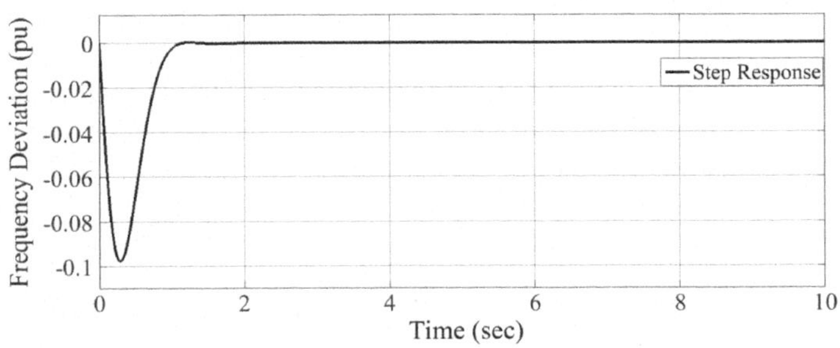

Fig. 8 GA-PID-based step response for 10% load change (case 3)

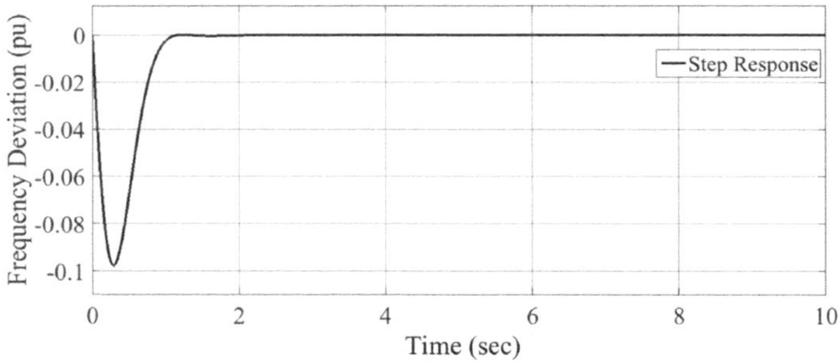

Fig. 9 PSO-PID-based step response for 10% load change (case 3)

Table 3 Comparison of GA-PID and PSO-PID for 10% load change (case 3)

	K_P	K_I	K_D	ITAE	Overshoot	Undershoot	Settling Time
GA-PID	1.2544	2.0000	0.2679	0.01913	0.00027431	-0.097744	1.30 Sec
PSO-PID	1.2573	2.0000	0.2680	0.01913	0.0001744	-0.09769	1.28 Sec

The performance index, overshoot, undershoot and settling time using GA-PID are 0.01913, 0.00027431, −0.097744 and 1.30 s, respectively, for 10% change in load in the system.

The performance index, overshoot, undershoot and settling time using PSO-PID are 0.01913, 0.0001744, −0.09769 and 1.28 s, respectively, for 10% change in load in the system.

The optimal value of K_P, K_I and K_D, corresponding values of ITAE, overshoot, undershoot and settling time using GA-PID and PSO-PID are shown in Table 3 for 10% change in load, respectively.

4.4 Case 4: Objective Function-ITSE

In this case, performance index ITSE is the objective function which is subjected to minimization. The step response of the system for 10% load change using GA-PID controller and PSO-PID is shown in Figs. 10 and 11, respectively.

The performance index, overshoot, undershoot and settling time using GA-PID are 0.0007085, 0, −0.074182 and 4.9 s, respectively, for 10% change in load in the system.

The performance index, overshoot, undershoot and settling time using PSO-PID are 0.0007047, 0, −0.074182 and 4.8 s, respectively, for 10% change in load in the system.

The optimal value of K_P, K_I and K_D, corresponding values of ITSE, overshoot, undershoot and settling time using GA-PID and PSO-PID are shown in Table 4 for 10% change in load, respectively.

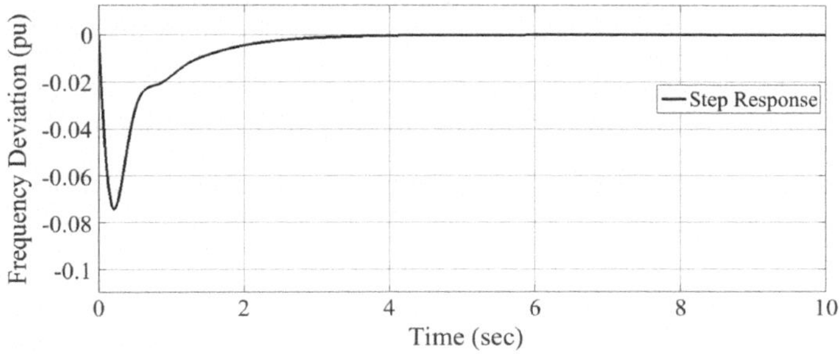

Fig. 10 GA-PID-based step response for 10% load change (case 4)

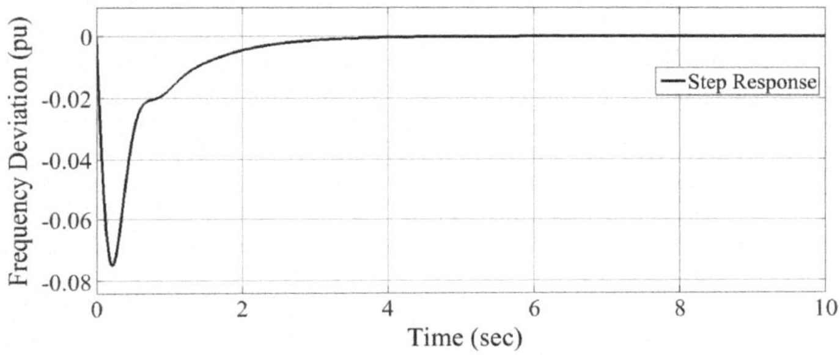

Fig. 11 PSO-PID-based step response for 10% load change (case 4)

Table 4 Comparison of GA-PID and PSO-PID for 10% load change (case 4)

	K_P	K_I	K_D	ITSE	Overshoot	Undershoot	Settling Time
GA-PID	1.9866	1.9992	0.4458	0.0007085	0	-0.074985	4.9 Sec
PSO-PID	2.0000	2.0000	0.4328	0.0007047	0	-0.074182	4.8 Sec

5 Conclusion

In this chapter, PSO and GA techniques are applied for designing of PID controller for LFC problem in single area power network. In place of considering one or two criteria, all the four performance indices, namely IAE, ISE, ITAE and ITSE, have been considered as the objective functions for solving LFC problem.

The results are analysed and compared based on value of performance indices, overshoot, undershoot and settling time of the step response of system.

From the results, it can be analysed and validated that the PSO technique-based PID controller gives better performance considering all the four performance indices, namely, ISE, IAE, ITSE and ITAE, as compared to GA technique-based PID controller.

Acknowledgements The authors are highly acknowledge and thankful to the Ministry of Human Resource Development (MHRD), New Delhi, for providing the financial assistance under TEQIP-III and to All India Council for Technical Education (AICTE), New Delhi, for providing financial assistance under the RPS Project File No. 8/36/RIFD/RPS/POLICY-I/2016–17, dated 02/August/2018.

The authors are also thankful to the Director, MITS, Gwalior, India, for providing the facilities for this research work.

References

1. Woodand AJ, Wollenberg BF (1996) Power generation, operation and control, 2nd edn. Wiley, New York, USA
2. Kothari DP, Nagrath IJ (2003) Modern power system analysis, 3rd edn. Tata McGraw Hill, New Delhi, India
3. Khezri R et al (2019) An intelligent coordinator design for GCSC and AGC in a two-area hybrid power system. Appl Soft Comput J 76:491–504
4. Bevrani H (2014) Robust power system frequency control, 2nd edn. Springer, New York, USA
5. Golshannavaz S, Khezri R, Esmaeeli M, Siano PL (2018) A two-stage robust intelligent controller design for efficient LFC based on Kharitonov theorem and fuzzy logic. J Ambient Intell Hum Comput 9:1445–1454
6. Bevrani H, Hiyama T (2011) Intelligent automatic generation control, CRC, New York, USA
7. Mohanty B, Panda S, Hota PK (2014) Controller parameters tuning of differential evolution algorithm and its application to load frequency control of multi-source power system. Int J Electr Power Energy Syst 54:77–85
8. Ibraheem, Kumar P, Kothari DP (2005) Recent philosophies of automatic generation control strategies in power systems. IEEE Trans Power System 20:46–57
9. Chaturvedi DK, Satsangi PS, Kalra PK (1999) Load frequency control: a generalized neural network approach. Int J Electr Power Energy Syst 21:405–415
10. Ghosal SP (2004) Optimization of PID gains by PSO in fuzzy based automatic generation control. Electr Power Syst Res 72:203–212
11. Parmar KPS, Majhi S, Kothari DP (2012) Load frequency control of a realistic power system with multi-source power generation. Int J Electr Power Energy Syst 42:426–433
12. Duman S, Yorukeren N, Altas IH (2012) Load frequency control of a single area power system using gravitational search algorithm. In: International symposium on innovations in intelligent systems and applications, pp 1–5
13. Nath V, Sambariya DK (2016) Application of NARMA L2 controller for load frequency control of multi-area power system. In. IEEE proceeding of 10th international conference on intelligent systems and control (ISCO 2016), vol 2, pp 352–358
14. Tan W (2010) Unified tuning of PID load frequency controller for power systems via IMC. IEEE Trans Power Syst 25:341–350
15. Sambariya DK, Fagna R (2017) A novel Elephant Herding Optimization based PID controller design for load frequency control in power system. In: International conference on computer, communications and electronics (Comptelix), vol 1, pp 595–600
16. Bhatt P, Roy R, Ghoshal SP (2010) GA/particle swarm intelligence based optimization of two specific varieties of controller devices applied to two-area multi-units automatic generation control 32(4):299–310
17. Ali ES, Abd-Elazim SM (2013) BFOA based design of PID controller for two area load frequency control with nonlinearities. Int J Electr Power Energy Syst 51:224–231
18. Shariq A, Bayyavarapu S (2015) Study of PID controllers to load frequency control systems with various turbine models. Int J Eng Res Appl 5:1(Part 5):92–98

Chapter 5
Combined Economic Emission Dispatch of Hybrid Thermal PV System Using Artificial Bee Colony Optimization

Salil Madhav Dubey, Hari Mohan Dubey, and Manjaree Pandit

Abstract Economical and reliable provision of electricity has been one of the most significant research objectives since decades. With time, various economic load dispatch (ELD) techniques have emerged in power market. Apart from using these methods, changes in the use of conventional source of energy and incorporating non-conventional sources have emerged in recent years. Solar photovoltaic (PV) generation helps reducing emissions and dependency on fossil fuels. This chapter presents combined economic emission dispatch (CEED) of a hybrid thermal solar PV system. Artificial bee colony (ABC) algorithm is used as optimization tool for the scenario involving six thermal plants and thirteen solar plants. The effectiveness of this method is compared and validated with other methods available in recent literature.

Keywords Economic load dispatch · Economic emission dispatch · Artificial bee colony algorithm · Solar PV system

Nomenclature

Min C	Objective function
$F_i(P_i)$	Fuel cost (in \$/h) for i-th power generating unit
$E_i(P_i)$	Emission (in kg/h) for i-th power generating unit
w	Weight ratio
ppf	Price penalty factor

S. M. Dubey (✉) · H. M. Dubey · M. Pandit
Madhav Institute of Technology and Science, MITS, Gwalior, Madhya Pradesh, India
e-mail: salil.dubey3107@gmail.com

H. M. Dubey
e-mail: harimohandubeymits@gmail.com

M. Pandit
e-mail: drmanjareep@gmail.com

M. Pandit et al. (eds.), *Nature Inspired Optimization for Electrical Power System*, Algorithms for Intelligent Systems,
https://doi.org/10.1007/978-981-15-4004-2_5

P_i	Power generated by i-th source
P_L	Power loss
P_d	Power demand at that instant
k_r	Penalty cost factor underestimation
$P_{i\,\min}$, $P_{i\,\max}$	Minimum and maximum power limits for i-th thermal generating source, respectively
u_i, l_i	Upper and lower bound of the solution space of objective function
rand (0,1)	A random number ∈ (0,1)
x_k	Randomly selected food source
φ_{mi}	Random number ∈ (−1,1)
P_m	Probability function
a_i, b_i, c_i	Fuel cost coefficients of i-th generating unit
$\alpha_i, \beta_i, \gamma_i$	Emission coefficients of i-th generating unit
P_{rated}	Rated output of a solar plant (MW)
T_{ref}	Reference temperature (25 °C in this case)
T_{amb}	Ambient temperature of solar plant
μ	Temperature coefficient of solar plant (–0.47%)
G_i	Incident solar radiation (W/m^2) at i-th hour
C_j	Cost per unit for j-th solar plant
Psch_j	Scheduled power for j-th solar plant
k_p	Penalty cost factor for overestimation
$f_m(x_m)$	Objective function value of x_m
$\text{fit}(x_m)$	Fitness of x_m
x_m	Initial food sources
v_{mi}	Neighbor food source
x_{mn}	New solution

1 Introduction

In twenty-first century, electricity has been an integral part of our lives. As per Global Energy & CO_2 Status (GECO) Report by International Energy Agency (IEA), high demand of electricity has increased the power demand by 4% or 900 TWh in 2018. This increasing demand was mostly met by thermal plants using coal or gas as major firing fuel. These thermal plants emit CO_2 which has reached up to 13 GT. In 2018, India's power demand increased by around 65 TWh or by 5.4% than previous year. To meet it efficiently, it is important that other than conventional sources of electricity, non-conventional sources like solar, wind, biomass and others should be encouraged. Solar is, however, mostly popular and easy to use option for harnessing renewable sources of energy. National Solar Mission aims to produce 100 GW of India's total power demand by 2022. Solar energy units also help in reducing carbon credits by producing less emissions. Therefore, incorporating solar PV units with thermal units can be an economical as well as environment-friendly option for power companies.

However, the distribution of load over the generating sources for an efficient and economical production is a main concern of power producing companies. Thus, ELD is a crucial step for efficient operation of power industry. The objective function of practical ELD problem is quite complex due to nonlinearity, multimodel and discontinuity associated with it. Due to this fact, its solution with traditional approach is not possible.

Considering Kyoto protocol of green energy policy, objective function for emission reduction is also combined with ELD. When the generating sources aim to produce energy with both minimum fuel costs and minimum emission levels simultaneously, the optimal solution problem will become a combined economic and emission dispatch (CEED) problem. As a solution for these types of problems, nature-inspired algorithms (NIAs) have become more popular among researchers since last decade because of the capability to provide near global minima solution for any type of complex optimization problem. The detailed review of NIA for solution of economic/emission dispatch problems can be found in [1, 2].

Nowadays, integration of renewable energy resources like wind, PV system, biomass, etc., has become more popular in order to reduce operational cost and emission [2–7]. Integration of uncertain renewable energy resources with conventional power plants complicates the objective function. For solution of these types of problems, a robust optimization approach is required.

In this chapter, artificial bee colony (ABC), inspired by foraging behavior of honey bees, is utilized for CEED solution of a hybrid thermal PV system. In solar PV generation, the output is variable as it depends on irradiance. Therefore, a probabilistic model is used for photovoltaic (PV) generation which takes into account the overestimation and underestimation cost factors [4].

The main objective of this chapter is to analyze the impact of renewable integration on operational fossil fuel cost and emission. This chapter is organized as follows: Problem formulation of this system is given in Sect. 2, the working of the optimization method is described in Sect. 3, results and discussion after using this model are explained in Sect. 4, and the conclusions drawn are compiled in Sect. 5.

2 Problem Formulation

2.1 Objective Function

The objective function for combined economic emission dispatch for a conventional thermal plant is given as [3]:

$$\text{Min}\, C = \sum_{i=1}^{6} (w \times F_i(P_i) + (1 - w) \times \text{ppf} \times E_i(P_i)) \tag{1}$$

ppf is given as

$$\text{ppf} = \frac{a_i \times P_{i\max}^2 + b_i \times P_{i\max} + c_i}{\alpha_i \times P_{i\max}^2 + \beta_i \times P_{i\max} + \gamma_i} \tag{2}$$

$$F_i(P_i) = a_i \times P_i^2 + b_i \times P_i + c_i(\$/\text{h}) \tag{3}$$

$$E_i(P_i) = \alpha_i \times P_i^2 + \beta_i \times P_i + \gamma_i(\text{kg/h}) \tag{4}$$

The power generated by solar plants is given by [4]:

$$\text{Ps}_j = P_{\text{rated}}\{1 + (T_{\text{amb}} - T_{\text{ref}}) \times \mu\} \times (G_i/1000)(\text{MW}) \tag{5}$$

After considering the overestimation and underestimation of solar cost, the total solar cost of operation becomes

$$\sum_{j=1}^{13} \text{Ps}_j \times C_j + \sum_{j=1}^{13} k_p \times (\text{Ps}_j - \text{Psch}_j) + \sum_{j=1}^{13} k_r \times (\text{Psch}_j - \text{Ps}_j) \tag{6}$$

$$\begin{aligned} \text{Min}C = & \sum_{i=1}^{6} (w \times F_i(P_i) + (1 - w) \times \text{ppf} \times E_i(P_i)) + \sum_{j=1}^{13} \text{Ps}_j \times C_j \\ & + \sum_{j=1}^{13} k_p \times (\text{Ps}_j - \text{Psch}_j) + \sum_{j=1}^{13} k_r \times (\text{Psch}_j - \text{Ps}_j) \end{aligned} \tag{7}$$

2.2 Equality Constraint

$$\sum_{i=1}^{6} P_i - P_L - P_d - \sum_{j=1}^{13} Ps_j = 0 \tag{8}$$

2.3 Inequality Constraint

$$P_{i\min} \le P_i \le P_{i\max} \tag{9}$$

3 Artificial Bee Colony Optimization

Artificial bee colony (ABC) optimization is swarm intelligence-based nature-inspired algorithm. It is based on the intelligent foraging behavior of honey bees proposed by Karaboga [8]. ABC algorithm is basically a population-based search in which locations called 'food sources' are modified by artificial bees from time to time. The bee's aim is to select the food sources with high amount of nectar and finally settle down at the location with the highest amount of nectar. There are three types of bees depending on their respective jobs in the swarm, namely employed bees, on-looker bees and scout bees. It is assumed that the number of employed bees is equal to the number of food sources (locations). The employed bees first calculate the nectar amount of the sources (fitness values) and then perform the waggle dance to transfer this information to the hive. On-looker bees identify and choose the high nectar food sources after observing the waggle dance pattern. After analyzing all locations and going through multiple iterations, some food sources are abandoned when they fail to converge to a point and do not improve in specified number of 'trail limit.' Then, the employed bees become scout bees and search for another food source in order to increase accuracy of the algorithm.

The analytical model of ABC algorithm is described as below:

Step 1: The initial food sources are randomly selected within upper and lower limit as

$$x_m = l_i + \text{rand}(0, 1) \times (u_i - l_i) \tag{10}$$

where u_i and l_i are the upper and lower bound of the solution space of objective function, rand (0, 1) is a random number within the range [0, 1].

Step 2: The neighbor food source v_{mi} is determined and calculated as

$$v_{mi} = x_{mi} + \phi_{mi}(x_{mi} - x_{ki}) \tag{11}$$

The fitness value is calculated by using (12) and (13), and then, greedy selection is applied between x_m and v_m.

$$\text{fit}(x_m) = \frac{1}{(f_m(x_m))}, \; f_m(x_m) > 0 \tag{12}$$

$$\text{fit}(x_m) = 1 + |f_m(x_m)|, \; f_m(x_m) < 0 \tag{13}$$

where f_m (x_m) is the objective function value of x_m.

Step 3: The quantity of a food source is evaluated by its probability (P_m) and the probability of all food sources. P_m is determined as

$$P_m = \frac{\text{fit}(x_m)}{\sum \text{fit}(x_m)} \tag{14}$$

where fit(x_m) is the fitness of x_m. On-looker bees search the neighborhoods of food source according to the expression.

$$v_{mi} = x_{mi} + \phi_{mi}(x_{mi} - x_{ki}) \quad (15)$$

Step 4: The new solutions are randomly searched by the scout bees. The new solution x_m will be discovered by the scout bees by using the expression.

$$x_m = l_i + \text{rand}(0, 1) \times (u_i - l_i) \quad (16)$$

4 Results and Discussion

4.1 Description of Test Cases

Case 1 This test system contains six thermal power units; its fuel cost, minimum and maximum power limits and emission coefficients are adapted from [3] and listed in Table 1.

Case 2 It is a hybrid test case having six thermal units similar to Case 1 and thirteen solar PV unit system. The required data of the solar PV units are adapted from [4] and also listed in Table 2 and Fig. 1. Table 2 gives data of power ratings and unit price for solar plants, and Fig. 1 represents solar radiance and temperature over 24 h of a particular day.

These systems were analyzed for six different load demand.

Table 1 Data for thermal units

Units	a (\$/MW^{2}h)	b (\$/MWh)	c (\$/h)	P_{min} (MW)	P_{max} (MW)	α (kg/MW^{2}h)	β (kg/MWh)	γ (kg/h)
1	0.15247	38.539	756.790	10	125	0.00419	0.32767	13.859
2	0.10587	46.159	451.320	10	150	0.00419	0.32767	13.859
3	0.02803	40.396	1049.99	35	250	0.00683	−0.54551	40.266
4	0.03546	38.305	1243.53	35	210	0.00683	−0.54551	40.266
5	0.02111	36.327	1658.56	135	325	0.00461	−0.51116	42.895
6	0.01799	38.270	1356.65	125	315	0.00461	−0.51116	42.895

Table 2 Data for solar units

Unit	1	2	3	4	5	6	7	8	9	10	11	12	13
P_{rated}	20	25	25	30	30	35	35	40	40	40	40	40	40
UR	0.22	0.23	0.23	0.24	0.24	0.25	0.26	0.27	0.275	0.28	0.28	0.28	0.28

Unit Rate (UR) in $/KWh

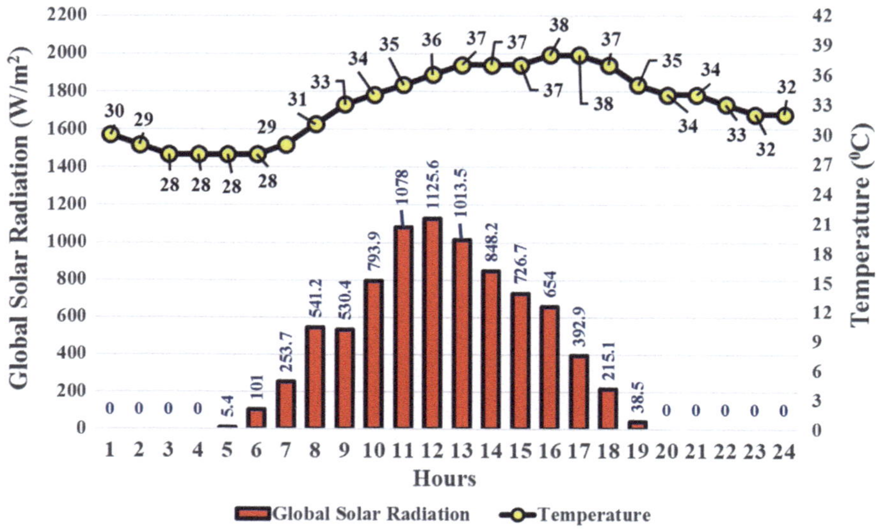

Fig. 1 Solar radiance and temperature data for solar PV units

4.2 Simulation Results

ABC is implemented for solution of ELD, EED and CEED as objective function for above two cases in MATLAB R2013a environment. The parameter used for simulation was considered as colony size: 50, limit: 100 and maximum number of iteration of 200 as stopping criteria. For each test case, ABC algorithm was run for 30 times for statistical analysis.

The optimal generation scheduling for Case 1 is for best cost solution, best emission solution and combined economic emission dispatch for different load demands are listed in Tables 3, 5 and 6, respectively. The statistical comparisons of results obtained by ABC are compared with the most recently reported methods such as

Table 3 Optimal generation scheduling obtained by ABC for ELD (Case 1)

O/P	PD:1150 MW	PD:1201 MW	PD:1235 MW	PD:1190 MW	PD:1251 MW	PD:1263 MW
P1	47.2767	56.1376	70.0711	51.9191	76.6280	81.5457
P2	32.1012	44.8624	64.9289	38.7872	74.3720	81.4543
P3	224.0408	250.0000	250.0000	249.2937	250.0000	250.0000
P4	206.5813	210.0000	210.0000	210.0000	210.0000	210.0000
P5	325.0000	325.0000	325.0000	325.0000	325.0000	325.0000
P6	315.0000	315.0000	315.0000	315.0000	315.0000	315.0000
TC	58,029.9316	60,779.2170	62,743.8311	60,174.4908	63,718.3422	64,470.2200
ES	1250.3462	1343.1350	1370.8751	1333.6995	1385.6598	1397.4752

Total Cost (TC) in $/h, Emission (ES) in kg/h

quadratic constraint programming (QCP) [3], genetic algorithm (GA) [3] and particle swarm optimization (PSO) [3] and are tabulated in Table 4. Here, it is observed that the optimum results in terms of minimum cost obtained by ABC are found to be lower than all previously reported methods.

The smooth cost convergence curve obtained by ABC algorithm for distinct power demand is presented in Fig. 2.

Similarly, for hybrid thermal PV system as in Case 2, best cost solution, best emission solution and combined economic emission dispatch for different load demands are listed in Tables 7, 8 and 9.

After the comparison of results for thermal system (Case 1) with hybrid thermal PV system (Case 2), it is clearly observed that significant reduction in total operational cost and emission can be achieved by integration of PV system. Considering the optimal generation schedule obtained using ABC as listed in Tables 3, 5, 6 and

Table 4 Statistical comparison of results (ELD Case 1)

Method	ABC	QCP [4]	GA [4]	PSO [4]
PD: 1150 MW				
Min Cost ($/h)	58,029.9316	60.438×10^3	61.434×10^3	60.335×10^3
Emission (kg/h)	1250.3462	1245.8	1130.5	1130.8
SD	0.00	NA	NA	NA
PD: 1201 MW				
Min Cost ($/h)	60,779.2170	63.909×10^3	63.883×10^3	62.270×10^3
Emission (kg/h)	1343.1350	1304.6	1237.4	1238.7
SD	0.00	NA	NA	NA
PD: 1235 MW				
Min Cost ($/h)	62,743.8311	64.280×10^3	65.461×10^3	64.508×10^3
Emission (kg/h)	1370.8751	1334.8	1260.3	1245.8
SD	0.00	NA	NA	NA
PD: 1190 MW				
Min Cost ($/h)	60,174.4908	63.566×10^3	63.626×10^3	61.367×10^3
Emission (kg/h)	1333.6995	1280.5	1245.7	1219.2
SD	0.00	NA	NA	NA
PD: 1251 MW				
Min Cost ($/h)	63,718.3422	66.000×10^3	66.000×10^3	65.349×10^3
Emission (kg/h)	1385.6598	1363.9	1302.6	1301.8
SD	0.00	NA	NA	NA
PD: 1263 MW				
Min Cost ($/h)	64,470.2200	68.180×10^3	67.043×10^3	66.919×10^3
Emission (kg/h)	1397.4752	1338.0	1332.6	1332.6
SD	0.00	NA	NA	NA

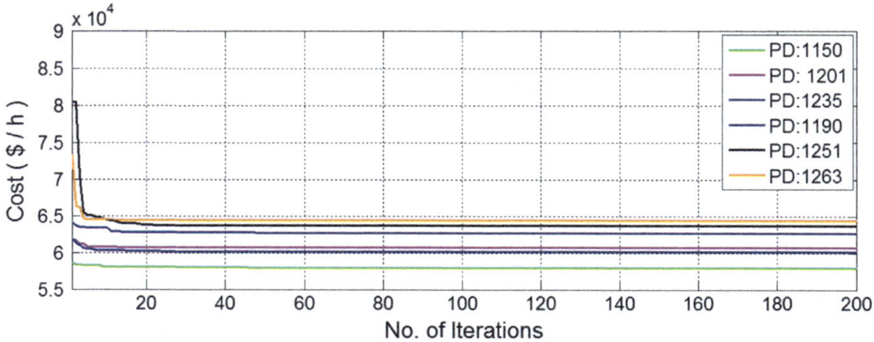

Fig. 2 Cost convergence curve for ELD obtained by ABC algorithm (Case 1)

Table 5 Optimal generation scheduling obtained by ABC for EED (Case 1)

O/P	PD:1150 MW	PD:1201 MW	PD:1235 MW	PD:1190 MW	PD:1251 MW	PD:1263 MW
P1	124.9546	124.9624	124.8797	124.8983	124.9611	124.9936
P2	145.8948	147.9542	150.0000	149.7836	149.9058	149.9992
P3	179.8940	172.8499	195.7672	186.8872	188.1472	195.9449
P4	187.0779	192.5855	197.4114	182.3255	175.2457	198.1936
P5	255.2884	288.7442	281.1173	283.1060	265.7937	299.5235
P6	256.8903	273.9037	285.8243	262.9994	285.9465	294.3453
TC	60,949.7171	63,453.5339	65,227.7514	62,974.3908	62,967.1920	66,595.7756
ES	1040.1455	1141.2559	1208.3131	1116.8091	1117.5838	1268.8282

Table 6 CEED solutions obtained by ABC algorithm (Case 1)

O/P	PD:1150 MW	PD:1201 MW	PD:1235 MW	PD:1190 MW	PD:1251 MW	PD:1263 MW
P1	124.9988	124.9825	125	124.9827	124.99	125
P2	130.9449	150	150	150	149.3567	150
P3	178.1171	190.9985	197.5969	190.7457	195.3109	202.2525
P4	180.2342	188.1318	193.1487	177.7272	199.5561	199.2559
P5	268.4053	270.9961	287.8487	274.3882	288.4889	291.7856
P6	267.2998	275.891	281.4033	272.1561	293.2975	294.7037
$\sum P$	1150.0001	1200.9999	1234.9976	1189.9999	1251.0001	1262.9977
Th. C	60,502.0131	63,542.7922	65,219.8169	62,978.8756	65,995.3138	66,617.0333
PPF	74.16	74.16	74.16	74.16	83.77	83.77
ES	1045.951	1137.8531	1208.2652	1116.2762	1242.8422	1268.5089

Thermal cost (Th. C) in $/h, price penalty factor (PPF)

Tables 7, 8 and 9, it is clearly observed that all the operational constraints are fully satisfied.

Table 7 Optimal generation scheduling for ELD obtained by ABC (Case 2)

O/P	PD:1150 MW	PD:1201 MW	PD:1235 MW	PD:1190 MW	PD:1251 MW	PD:1263 MW
P1	29.2126	26.5713	27.0916	27.1928	32.1476	34.4316
P2	10.0000	10.0000	10.0000	10.0000	10.3129	13.6023
P3	125.7806	111.4133	114.2436	114.7942	141.7456	154.1699
P4	128.9097	117.5529	119.7901	120.2254	141.5296	151.3506
P5	263.3820	244.3051	248.0631	248.7943	284.5805	301.0776
P6	255.0695	232.6840	237.0938	237.9518	279.9444	299.3025
TS	812.3543	742.5266	756.2823	758.9584	890.2605	953.9345
SS	337.6457	458.4734	478.7177	431.0416	360.7395	309.0655
FC	41,260.2801	37,975.2279	38,617.9188	38,743.2057	44,991.7518	48,092.1040
SC	62.3315	84.6370	88.3742	79.5729	66.5947	57.0554
TC	41,322.6116	38,059.8649	38,706.293	38,822.7786	45,058.3465	48,149.1594
ES	648.2185	544.5864	564.0036	567.8381	778.3028	891.6221

Thermal Share (TS) in MW, Solar Share (SS) in MW, Fuel Cost (FC) in $/h, Solar Cost (SC) in $/h

Table 8 Optimal generation scheduling for EED obtained by ABC (Case 2)

O/P	PD:1150 MW	PD:1201 MW	PD:1235 MW	PD:1190 MW	PD:1251 MW	PD:1263 MW
P1	107.5481	85.7773	88.5040	89.0345	115.0621	124.9108
P2	107.5481	85.7773	88.5040	89.0345	115.0621	130.1043
P3	129.8999	116.5442	118.2170	118.5424	134.5096	141.9673
P4	129.8999	116.5442	118.2170	118.5424	134.5096	140.5491
P5	188.7292	168.9418	171.4202	171.9023	195.5586	204.5016
P6	188.7292	168.9418	171.4202	171.9023	195.5586	211.9016
TS	852.3543	742.5266	756.2823	758.9584	890.2605	953.9345
SS	337.6457	458.4734	478.7177	431.0416	360.7395	309.0655
FC	45,379.4121	39,435.2139	40,163.3805	40,305.5864	47,500.2194	51,119.8638
SC	62.3315	84.6370	88.3742	79.5729	66.5947	57.0554
TC	45,441.7436	39,519.8509	40,251.7547	40,385.1593	47,566.8141	51,176.9192
ES	585.6901	460.7387	475.2910	478.1586	633.4673	719.3105

Table 9 CEED solutions obtained by ABC algorithm for thermal PV system (Case 2)

O/P	PD:1150 MW	PD:1201 MW	PD:1235 MW	PD:1190 MW	PD:1251 MW	PD:1263 MW
P1	76.7963	66.8135	68.78	69.1626	90.3991	99.8163
P2	76.0605	64.5959	66.8543	67.2937	91.5655	102.2532
P3	131.9177	122.116	124.0469	124.4225	142.1001	150.9211
P4	132.0922	122.4848	124.3774	124.7456	142.0758	150.7431
P5	198.4568	184.0622	186.8979	187.4495	212.6069	225.5754
P6	197.0308	182.4543	185.3258	185.8845	211.5131	224.6253
$\sum P$	812.3543	742.5267	756.2823	758.9584	890.2605	953.9344
PV share	337.6457	458.4733	478.7177	431.0416	360.7395	309.0655
Th. C	42,273.2664	38,753.0045	39,439.6336	39,573.6053	46,401.6846	49,796.9395
PV cost	62.3315	84.637	88.3742	79.5729	66.5947	57.0554
PPF	49.66	49.66	49.66	49.66	57.178	57.17
ES	545.4643	466.4741	481.3721	484.3082	641.627	729.0764
TC	42,335.5979	38,837.6415	39,528.0078	39,653.1782	46,468.2793	49,853.9949

5 Conclusion

In this chapter, an efficient ABC algorithm is implemented for optimal generation scheduling of hybrid thermal PV system. Probabilistic modeling for solar PV is incorporated for computing the solar power output.

By analyzing the results presented in tabular form as listed in above section, various conclusions have been drawn, which are given below.

In Case 1, with increase in load demand, the fuel cost for thermal units also increases. From Table 4, it can be observed that the fuel cost for operation of thermal plants is lowest when ABC algorithm is used for ELD. Thus, ABC proves to be more economical than QCP, GA and PSO. There is a saving of 4–5% in fuel costs in Case 1 for ABC as compared to other algorithms.

Comparing results of Case 2 with Case 1, it is evident that incorporating solar PV units with thermal units is found to be more economical than using only thermal units. Here, it is clearly observed by Fig. 3 that the highest solar share is achieved when the global radiation is maximum among 24 h.

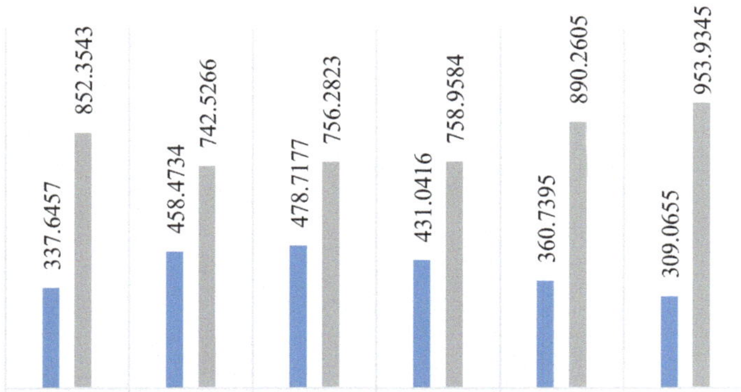

	1150	1201	1235	1190	1251	1263
Solar Share	337.6457	458.4734	478.7177	431.0416	360.7395	309.0655
Thermal Share	852.3543	742.5266	756.2823	758.9584	890.2605	953.9345

Fig. 3 Solar share and thermal share distribution over the load demands

Acknowledgements The authors acknowledge financial support provided by AICTE-RPS project File No. 8-36/RIFD/RPS/POLICY-1/2016-17 dated 2.9.2017 and TEQIP III. The authors also thank the Director and management of M.I.T.S. Gwalior, India, for providing facilities for carrying out this work.

References

1. Qu BY, Zhu YS, Jiao YC, Wu MY, Suganthan PN, Liang JJ (2018) A survey on multi-objective evolutionary algorithms for the solution of the environmental/economic dispatch problems. Swarm Evol Comput 38:1–11
2. Dubey HM, Pandit M, Panigrahi BK (2018) An overview and comparative analysis of recent bio-inspired optimization techniques for wind integrated multi-objective power dispatch. Swarm Evol Comput 38:12–34
3. Joshi PM, Verma HK (2019) An improved TLBO based economic dispatch of power generation through distributed energy resources considering environmental constraints. Sustain Energy Grids Netw. https://doi.org/10.1016/j.segan.2019.100207
4. Khan NA, Awan AB, Mahmood A, Razzaq S, Zafar A, Sidhu GAS (2015) Combined emission economic dispatch of power system including solar photo voltaic generation. Energy Convers Manage 92:82–91
5. Reddy SS, Momoh JA (2015) Realistic and transparent optimum scheduling strategy for hybrid power system. IEEE Trans Smart Grid 6(6):3114–3125
6. Das D, Bhattacharya A, Ray RN (2019) Dragonfly algorithm for solving probabilistic economic load dispatch problems 31(5):1–17
7. Zhang Y, Gatsis N, Giannakis GB (2013) Robust energy management for microgrids with high-penetration renewables. IEEE Trans Sustain Energy 4(4):944–953
8. Karaboga D, Basturk B (2007) A powerful and efficient algorithm for numerical function optimization: Artificial Bee Colony (ABC) algorithm. J Glob Optim 39(3):459–171

Chapter 6
Dynamic Scheduling of Energy Resources in Microgrid Using Grey Wolf Optimization

Salil Madhav Dubey, Hari Mohan Dubey, and Manjaree Pandit

Abstract Continuous and sustainable electricity is one of the major concerns in this modern world. This has led to the implementation of microgrid (MG) in order to establish an independent, efficient and cost-effective power supply system. The generation in MG can be conventional or non-conventional but due to increasing power demand, high fuel prices, scarcity of fossil fuels and degrading environment, there is a growing demand of using renewable energy sources (RS) for power generation. Solar PV units play an indispensable part in producing clean energy and coping with this modern-day power demand challenges. Grey wolf optimization (GWO), which is a metaheuristic technique inspired by the hierarchical hunting mechanism of grey wolves, is used in this chapter for solving a multi-objective problem in a dynamic environment of a microgrid. Dynamic dispatch is a more practical way which aims to provide an optimum solution in a scheduling horizon over twenty-four hours a day. A hybrid system comprising six conventional thermal plants and a solar farm containing thirteen solar PV units are discussed in this chapter. The performance and effectiveness of GWO are compared and validated with other two well-proven methods ABC and DE.

Keywords Microgrid · RS integration · GWO · Dynamic scheduling · Solar farm · Multi-objective scheduling

S. M. Dubey (✉) · H. M. Dubey · M. Pandit
Madhav Institute of Technology and Science, MITS, Gwalior, Madhya Pradesh, India
e-mail: salil.dubey3107@gmail.com

H. M. Dubey
e-mail: harimohandubeymits@gmail.com

M. Pandit
e-mail: drmanjareep@gmail.com

M. Pandit et al. (eds.), *Nature Inspired Optimization for Electrical Power System*, Algorithms for Intelligent Systems,
https://doi.org/10.1007/978-981-15-4004-2_6

Nomenclature

a_i, b_i, c_i	Fuel cost coefficients of i-th generating unit
P_i	Output power in MW of i-th generating unit
$\alpha_i, \beta_i, \gamma_i$	Emission coefficients of i-th generating unit
P_{rated}	Rated output of a solar plant
T_{ref}	Reference temperature taken (25 °C in this case)
T_{amb}	Ambient temperature of solar plant
μ	Temperature coefficient of solar plant (–0.50% in this case)
S_t	Incident solar radiation (W/m2) at t-th hour
P_L	Power loss
UR_i, DR_i	Up rate and down rate of ith generating unit, respectively
A, C	Coefficient vectors
$\mathcal{X}(t)$	Position vector of the prey
$\mathcal{X}$	Position vector of a grey wolf
$r1, r2$	Random vectors ∈[0, 1]
$\mathcal{X}_1, \mathcal{X}_2, \mathcal{X}_3$	Best position of alpha (α), beta (β) and delta (δ), respectively
$\mathcal{X}(t+1)$	Final position

1 Introduction

Sustainable, renewable, efficient and economical energy systems are the need of the hour for meeting the power demand of increased population. Implementation of microgrid (MG) has gained popularity as a solution to this increased power demand. However, MG has its own challenges for economic operations. Uncertainty in the output of renewable energy sources (RES), energy storage (ES) capacity management, optimization of MG operation with real-time electricity price in market, minimizing operational cost and emissions are some challenges faced when MG is incorporated in the power system [1]. Solutions to these problems like dynamic scheduling of MG using NSGA-II algorithm [2], use of approximate dynamic programming and deep recurrent neural network learning in MG energy management [3], short term generation scheduling [4], scheduling in a CHP-based MG for economic power sharing [5], etc., have evolved to fulfil the interests of all stakeholders in power market.

In recent years, a lot of researchers have been focusing on the operation of MG. Optimal scheduling has always been one of the most important functions in minimizing the net cost of MG [6]. Dynamic optimal scheduling is a good option for MG operation because it considers the lowest cost in scheduling as well as coordinates among different distribution generations (DERs) over many periods.

In India, more than 70% conventional sources of energy are thermal plants which use coal as major fuel. Burning of coal produces harmful gases which degrade our air quality. Also, the price of fuel used is increasing day by day. Under these conditions, sharing of demand by DERs is not only governed by the units' capability of minimizing the total fuel cost of system generation but also the capability of satisfying the emission requirements. Many optimization algorithms have been used for solving this problem of minimizing fuel cost and emissions. Metaheuristic optimization techniques have gained popularity within last two decades for solution of complex optimization problem. Grey wolf optimization (GWO) [7] is a recently developed metaheuristic technique which is inspired by the hierarchal arrangement in hunting mechanism of grey wolfs.

In this chapter, GWO is used for dynamic scheduling of energy resources considering environmental constraints. Remaining chapters are organized as follows: Problem formulation of this system is given in Sect. 2, the working of the optimization method is described in Sect. 3, results and discussion after using this model are explained in Sect. 4 and the conclusions drawn are compiled in Sect. 5.

2 Problem Formulation

The fuel costs of the conventional generators in a dynamic environment of 24 h which is a convex polynomial can be mathematically expressed as (in \$/h) [10]:

$$F(P) = \sum_{t=1}^{24} \sum_{i=1}^{6} \left\{ a_i \times P_i^2(t) + b_i \times P_i(t) + c_i \right\} \tag{1}$$

Similarly, emission dispatch function (in Kg/h) is also a convex polynomial and can be written as [10]:

$$E(P) = \sum_{t=1}^{24} \sum_{i=1}^{6} \left\{ \alpha_i \times P_i^2(t) + \beta_i \times P_i(t) + \gamma_i \right\} \tag{2}$$

Thus, the multi-objective economic emission dispatch problem can be mathematically stated as [10]:

$$C(P) = \sum_{t=1}^{24} \sum_{i=1}^{6} \left[\left\{ a_i P_i^2(t) + b_i P_i(t) + c_i \right\} + \mathrm{ppf} \times \left\{ \alpha_i \times P_i^2(t) + \beta_i \times P_i(t) + \gamma_i \right\} \right] \tag{3}$$

where ppf is price penalty factor which is given by

$$\mathrm{ppf} = \frac{\left\{ a_i P_{i\max}^2(t) + b_i P_{i\max}(t) + c_i \right\}}{\alpha_i \times P_i^2(t) + \beta_i \times P_i(t) + \gamma_i} \tag{4}$$

The power generated by each solar PV unit (in MW) at t-th hour in a solar farm is given by [11]:

$$P_{\text{gs}} = P_{\text{rated}}\left\{1 + \mu(T_{\text{amb}} - T_{\text{ref}}) \times \frac{S_t}{1000}\right\} \tag{5}$$

Cost of operation for the solar farm for 24 h is given as:

$$\sum_{t=1}^{24}\sum_{j=1}^{13} P_{\text{gs}} \times C_j \tag{6}$$

The multi-objective cost function of the hybrid system becomes [11]:

$$C(P) = \sum_{t=1}^{24}\left[w \times \left(\sum_{i=1}^{6}\left\{a_i P_i^2(t) + b_i P_i(t) + c_i\right\}\right) + \text{ppf} \times (1 - w) \times \left(\sum_{i=1}^{6}\left\{\alpha_i \times P_i^2(t) + \beta_i \times P_i(t) + \gamma_i\right\}\right) + \sum_{j=1}^{13} P_{\text{gs}} \times C_j\right] \tag{7}$$

2.1 *Inequality Constraints*

The power generated by the conventional thermal plants as well as the RS (Solar PV farm) must lie between maximum and minimum limits. Mathematically,

$$P_i^{\min} \le P_i \le P_i^{\max} \tag{8}$$

$$P_{\text{gs}}^{\min} \le P_{\text{gs}} \le P_{\text{gs}}^{\max} \tag{9}$$

The ramp rate limits for thermal unit power generation are considered in this problem. The power generation of thermal units is constrained by the ramp rate limits as follows:

$$P_i^t - P_i^{t-1} \le \text{UR}_i \tag{10}$$

$$P_i^{t-1} - P_i^t \le \text{DR}_i \tag{11}$$

2.2 *Equality Constraints*

The power generated at any instant of time by all the thermal plants and the RS (Solar PV farm) should satisfy the total desired load of the system which is mathematically described as:

$$P_{\text{Load}} = \sum_{i=}^{6} P_i + \sum_{j=1}^{13} P_{\text{gs}} + P_L \tag{12}$$

3 Grey Wolf Optimization

Grey wolf optimization (GWO) is belonging to the family of swarm intelligence [7]. Its analytical model mimics the intelligent, self-organized group behaviour of grey wolves for hunting prey in nature. Grey wolves live in a group of 5–15 members. They follow a proper hierarchy with four types of member represented as Alpha, Beta, Delta and Omega. The social hierarchy of grey wolves is illustrated in Fig. 1. Systematic organization and discipline are their main strength.

Group leader is male/female represented by Alpha. He or She is only the decision maker for hunting, walking and selection of place for sleeping. Beta wolf has second place in social hierarchy and helps group leader in decision making. Delta is the subordinates of alpha and beta but they dominate over omega. Delta has four subgroups: Scouts, Sentinels, Hunters and Caretakers. Scouts are responsible for watching boundary territory and warning the group members in case of any danger. Sentinels are responsible for the protection of group members. Hunters help alpha and beta in hunting and also responsible for arranging the food for the group members. Weak and wounded member are taken care by caretakers. Omega plays the role of scapegoat in the group and they generally eat at last only.

On the basis of above-disciplined group behaviour, the analytical model of GWO is described by three phases during hunting which are described as below.

(a) **Entrapment of prey**

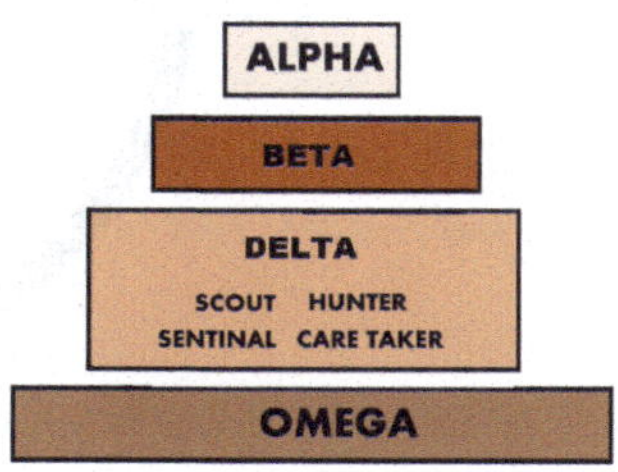

Fig. 1 Social hierarchy of grey wolves in nature

In its first phase, model is based upon assumption that grey wolves update their position one with respect to other in n-dimensional search space as below [7].

$$\mathcal{D} = |\mathcal{C} \times \mathcal{X}_P(t) - \mathcal{X}(t)| \tag{13}$$

$$\mathcal{X}(t+1) = \mathcal{X}_P(t) - \mathcal{A} \times \mathcal{D} \tag{14}$$

$$\mathcal{A} = 2ar_1 - a \tag{15}$$

$$\mathcal{C} = 2r_2 \tag{16}$$

$$a = 2 - (t)\left(\frac{2}{T}\right) \tag{17}$$

The value '*a*' is linearly decreased from 2 to 0 over the course of iterations and Fig. 2 illustrates this phase

(b) **Hunting of Prey**

In order to simulate self-organized and group behaviour of grey wolves, alpha, beta and gamma are considered as three best solutions. Alpha is assumed to be closest to the best solution followed by the solution of beta and gamma. Therefore, during optimization process, first three solutions are considered as the best and remainders are considered as omega. The position is updated with respect to the position of

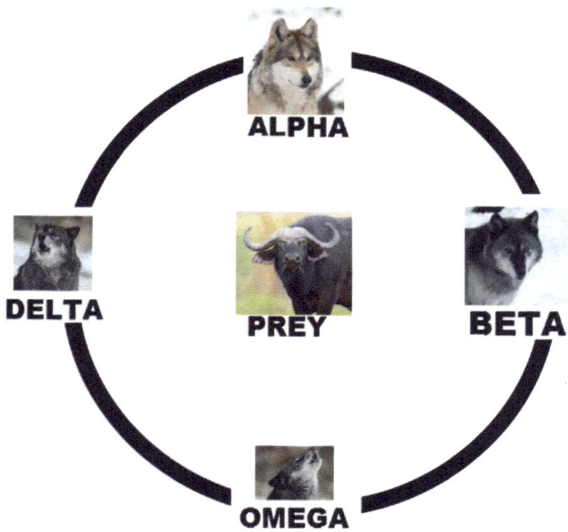

Fig. 2 Entrapment of prey phase

omega. The position of omega (ω) will vary as per the current best position in algorithm. The final position is defined with respect to position of alpha, beta and delta in search space as below.

$$\mathcal{D}_\alpha = |\mathcal{C}_1.\mathcal{X}_\alpha - \mathcal{X}|,\ \mathcal{D}_\beta = \left|\mathcal{C}_2.\mathcal{X}_\beta - \mathcal{X}\right|,\ \mathcal{D}_\delta = |\mathcal{C}_3.\mathcal{X}_\delta - \mathcal{X}| \tag{18}$$

$$\mathcal{X}_1 = \mathcal{X}_\alpha(t) - \mathcal{A}_1 \times \mathcal{D}_\alpha, \mathcal{X}_2 = \mathcal{X}_\beta(t) - \mathcal{A}_2 \times \mathcal{D}_\beta, \mathcal{X}_3 = \mathcal{X}_\delta(t) - \mathcal{A}_3 \times \mathcal{D}_\delta \tag{19}$$

$$\mathcal{X}(t+1) = \frac{1}{3} \times (\mathcal{X}_1 + \mathcal{X}_2 + \mathcal{X}_3) \tag{20}$$

(c) **Attacking the Prey**

In the last stage, grey wolf attacks the prey. In the analytical model, it can be realized by shrinking value of "*a*" from 2 to 0 as iteration progresses and hence $\mathcal{A}$ reduces. The last stage in hunting is attacking the prey when the prey has stopped. This can be achieved mathematically by reducing the value of a gradually from 2 to 0, consequently, $\mathcal{A}$ is varied randomly in range [−1, 1].

4 Results and Discussion

The main objective of this chapter is to find the impact of renewable integration on operating cost of fuel and quantity of emissions released, which is discussed in two cases. First case involving only thermal units and second case is a hybrid arrangement of thermal plants with solar PV integration.

4.1 Description of Test Cases

Case 1 This test system contains six thermal power units; its fuel cost, minimum and maximum power limits and emission coefficients which are adapted from [10] and listed in Table 1.

Case 2 It is a hybrid test case having six thermal units similar to Case 1 and a solar PV farm comprising of 13 PV units. The required data of the solar PV farm are adapted from [11] and illustrated in Fig. 3 and listed in Table 2. Figure 4 provides the data of temperature (°C) and solar radiation (W/m^2) of PV on a single day for 24 h. Table 2 gives data of rated power and per unit cost of thirteen PV units in the solar farm.

Table 1 Data related to six conventional thermal power plants

Unit	1	2	3	4	5	6
a_i ($/MW^{2}h)	0.007	0.0095	0.009	0.009	0.008	0.0075
b_i ($/MWh)	7	10	8	11	10.5	12
$c_i(\$/h)$	240	200	220	200	220	190
Pmin	100	50	80	50	50	50
Pmax	500	200	300	150	200	120
α_i(Kg/MW^{2}h)	0.00419	0.00419	0.00683	0.00683	0.00461	0.00461
β_i (Kg/MWh)	0.32767	0.32767	-0.54551	-0.54551	-0.51116	-0.51116
γ_i(Kg/h)	13.8593	13.8593	40.2669	40.2669	42.8955	42.8955
UR (MW/h)	80	50	65	50	50	50
DR(MW/h)	120	90	100	90	90	90

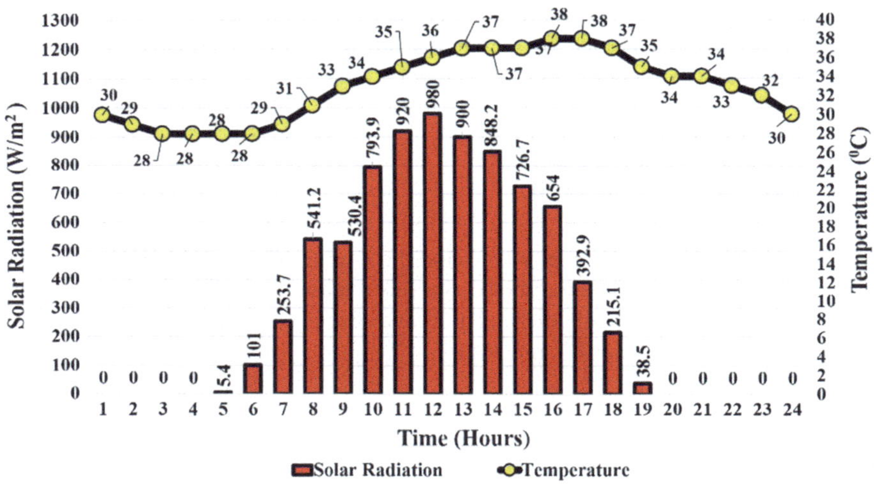

Fig. 3 Solar PV data of temperature (°C) and radiation (W/m^2)

4.2 *Simulation Results*

GWO is implemented for solution of ELD, EED and CEED problem in MATLAB R2013a environment. For each case, GWO algorithm was run for 30 times and best results are tabulated in Tables 3, 4, 5, 6 and 7.

The performance of GWO with recent methods like artificial bee colony (ABC) [8] and differential evolution (DE) [9] is given in Table 3. Table 4 tabulates the optimum scheduling of the six thermal units for CEED. The parameters considered in implementing the algorithms are given in Table 5. Here, it is observed that the optimum results in terms of minimum cost and least emissions obtained by GWO

Table 2 Rated power and per unit price of solar PV units in the solar farm

Unit	1	2	3	4	5	6	7	8	9	10	11	12	13
P_{rated}	20	25	25	30	30	35	35	40	40	40	40	40	40
UR	0.22	0.23	0.23	0.24	0.24	0.25	0.26	0.27	0.275	0.28	0.28	0.28	0.28

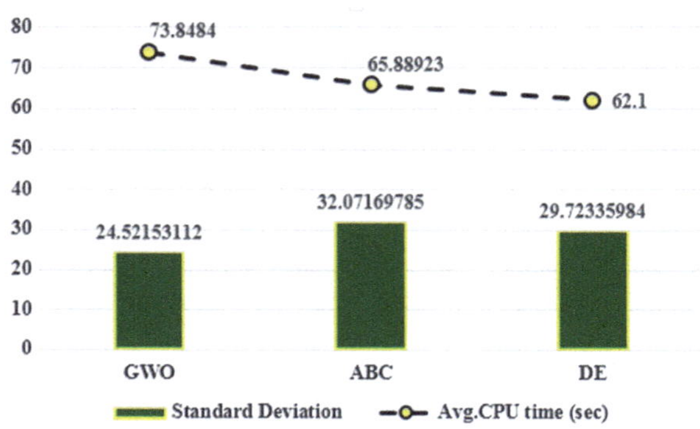

Fig. 4 Statistical and computational comparison of Case 1

Table 3 Comparison of cost and emissions for different algorithms (Case 1)

Method used	Thermal cost ($/h)	Total emission (Kg/h)
Economic load dispatch		
GWO	308039.0700	34101.0440
ABC	309591.6145	33901.0262
DE	308078.0118	33297.0680
Economic emission dispatch		
GWO	316174.2286	25577.1096
ABC	316172.2100	25579.5970
DE	316172.5736	25606.2910
Combined economic emission dispatch		
GWO	313504.6600	26594.1923
ABC	309361.6796	32703.8763
DE	309569.9413	33492.4277

are lowest as compared to the results obtained by simulation using ABC [8] and DE [9]. The statistical comparison in Fig. 4 illustrates that though the average CPU time in computation is more for GWO than ABC and DE, the standard deviation obtained in results by GWO is lowest than the other two methods.

The optimal solution in terms of cost and emission for hybrid thermal–PV system is listed in Table 6. By comparing results, it can be observed that the total cost for hybrid system is found to be lowest for GWO as compared to other two metaheuristic methods for all three objective functions taken into consideration. The optimal generation scheduled for CEED obtained using GWO is tabulated in Table 7. Here, it is observed that all associated operational constraints (8)–(11) are fully satisfied.

Table 4 Optimum generation schedule (CEED) obtained by GWO (Case 1)

Hour	P1	P2	P3	P4	P5	P6	Load (MW)
1	259.5374	149.3329	125.9369	109.1559	195.7285	115.3084	955
2	208.8734	123.3427	147.2819	148.8409	196.8753	116.7858	942
3	181.6658	155.7335	200.0944	131.9412	168.2502	115.3149	953
4	224.8678	102.1668	145.7005	143.4121	197.0021	116.8507	930
5	236.0181	138.2135	163.1357	105.9371	173.6471	118.0485	935
6	252.2066	120.3990	197.5179	148.2297	146.3651	98.2817	963
7	277.2475	125.9615	145.4991	148.2154	176.7064	115.3701	989
8	266.3530	148.9245	199.5718	122.3128	170.4414	115.3965	1023
9	310.7137	194.4492	153.4075	149.3816	199.5837	118.4643	1126
10	281.0990	198.1337	204.9100	148.3168	199.5668	117.9737	1150
11	350.3692	166.2672	219.3761	146.1391	198.8513	119.9971	1201
12	335.4002	175.6566	260.9748	149.5477	197.1420	116.2786	1235
13	369.6719	170.7951	187.6973	147.0773	197.3182	117.4401	1190
14	348.9233	198.5079	238.5792	148.1412	197.9208	118.9275	1251
15	404.2717	196.3962	237.2358	113.5441	192.9023	118.6499	1263
16	354.7831	198.6413	242.3785	146.6850	190.4917	117.0203	1250
17	394.5981	157.5716	203.4344	149.6969	198.7658	116.9333	1221
18	319.0318	197.9201	260.4055	109.1597	197.6255	117.8574	1202
19	389.6461	159.2299	219.9534	122.7978	149.4363	117.9365	1159
20	307.3071	134.1719	221.9858	147.3151	164.8292	116.3909	1092
21	242.0499	160.1049	176.4222	149.5229	177.4241	117.4760	1023
22	243.2312	106.4994	171.2364	145.3539	198.5802	119.0989	984
23	189.3753	153.2756	215.8568	148.1036	150.9817	117.4069	975
24	227.2540	166.6708	147.5936	148.4603	167.8862	102.1351	960

Table 5 Parameters used for different algorithms

Optimization	Population size (PS)	Food number	Limit	Max cycle	F1	F2	CR
GWO	100	–	–	100	–	–	–
ABC	100	PS/2	100	100	–	–	–
DE	150	–	–	100	0.2	0.2	0.8

5 Conclusion

This chapter focuses on using recently evolved nature-inspired technique named as grey wolf optimization (GWO) for solution of a hybrid thermal–PV system working

Table 6 Comparison of cost and emissions for different algorithms (Case 2)

Algorithms	Thermal cost ($/h)	PV cost ($/h)	Total cost ($/h)	Total emission (Kg/h)
Economic load dispatch				
GWO	268606.9160	856.4677	269463.3836	24453.7943
ABC	269214.9327	856.4677	270071.4004	24015.6247
DE	269492.3920	856.4677	270348.8597	24049.3635
Economic emission dispatch				
GWO	269135.3267	856.4677	269991.7944	23677.4826
ABC	269648.7819	856.4677	270505.2496	23756.6617
DE	269647.1498	856.4677	270503.6175	23759.0086
Combined economic emission dispatch				
GWO	269007.5291	856.4677	269955.7506	23939.7894
ABC	269451.6499	856.4677	270308.1176	23801.7434
DE	269648.1603	856.4677	270504.6279	23758.5261

as power producers in a microgrid in island mode. After analysing the illustrations above, it can be concluded that GWO provides better results as compared to two other well-proven optimization techniques which are ABC and DE. In dynamic environment, the GWO algorithm converged in an efficient manner for solution of environmental/economic dispatch problem in dynamic environment without violating any constraint.

In Case 1, GWO optimizes the minimal cost (ELD) and gives least emissions (EED) as compared to ABC and DE. In Case 2, the microgrid using thermal–PV units as DERs have lesser cost of operation, lower fuel cost and lesser emissions than in Case 1. Thus, using renewable sources of energy will economically and ecologically make the existing microgrid more efficient.

Microgrid using the proposed hybrid thermal–PV system implementing GWO as optimization methodology will be an economic and efficient way to solve the modern-day multi-objective power scheduling problems.

Table 7 Optimum generation schedule (CEED) obtained by GWO (Case 2)

Hour	P1	P2	P3	P4	P5	P6	PV output (MW)
1	291.6677	128.3787	226.496	96.4727	122.8774	89.1075	0
2	308.716	128.2172	200.563	102.6369	122.6878	79.1791	0
3	333.0985	117.9031	209.3775	95.4665	114.0431	83.1113	0
4	301.2931	127.0684	201.1006	83.5105	137.4587	79.5687	0
5	310.3517	137.8202	183.8572	94.0479	125.8481	80.7345	2.34036
6	287.9521	137.5316	198.416	103.5973	111.5374	80.1922	43.7734
7	285.7395	112.1214	168.3167	98.5323	131.7926	83.1021	109.39544
8	236.4054	77.2908	167.7723	95.6519	127.8353	87.0601	230.98416
9	288.7889	106.2189	183.7551	97.3922	146.7348	79.0691	224.04096
10	200.9361	111.4404	198.5657	91.0967	132.4942	81.8701	333.59678
11	213.1547	124.5768	200.5624	97.1991	86.8663	94.0807	384.5600
12	242.4135	117.6926	168.4934	99.1394	113.769	86.0081	407.4840
13	261.721	118.4019	150.4681	93.4949	112.4729	81.2012	372.24 00
14	272.2764	109.4945	205.7171	93.4321	143.2713	75.9931	350.81552
15	313.3497	129.3932	209.7549	99.4467	126.4515	84.041	300.56312
16	324.11	114.4542	194.7587	113.0236	137.8464	96.7515	269.0556
17	348.7288	138.4968	199.6722	109.6982	163.7378	99.0271	161.63906
18	358.7244	184.5096	237.4808	105.422	142.4219	84.4759	88.96536
19	437.3826	125.3974	207.0658	107.7161	185.8632	79.4819	16.0930
20	366.19	140.0392	243.4533	96.4065	149.4269	96.4841	0
21	361.6952	139.283	217.7123	111.0116	101.233	92.0649	0
22	324.2422	133.4168	191.4634	115.4422	134.5866	84.8488	0
23	333.5948	120.1099	219.0381	74.4675	148.1716	79.6181	0
24	304.34	124.8371	202.8644	99.884	134.9458	93.1287	0

Acknowledgement The authors acknowledge financial support provided by AICTE-RPS project File No. 8-36/RIFD/RPS/POLICY-1/2016-17 dated 2.9.2017 and TEQIP III. The authors also thank the Director and management of M.I.T.S. Gwalior, India, for providing facilities for carrying out this work.

References

1. Wang T, O'Neil D, Kamath H (2015) Dynamic control and optimization of distributed energy resources in a microgrid. IEEE Trans Smart Grid 6(6):2884–2894
2. Zhao F, Yuan J, Wang N (2019) Dynamic economic dispatch model of microgrid containing energy storage components based on a variant of NSGA-II algorithm. Energies 12(5):871–885

3. Zeng P, Li H, Haibo H, Li S (2019) Dynamic energy management of a microgrid using approximate dynamic programming and deep recurrent neural network learning. IEEE Trans Smart Grid 10(4):4435–4445
4. Logenthiran T, Srinivasan D (2009) Short term generation scheduling of a microgrid. TENCON. https://doi.org/10.1109/TENCON.2009.5396184
5. Basu AK, Bhattacharya A, Chowdhury S, Chowdhury SP (2012) Planned scheduling for economic power sharing in a CHP-based micro-grid. IEEE Trans Power Syst 27(1):30–38
6. Rezvani A, Gandomkar M, Izadbakhsh M, Ahmadi A (2015) Environmental/economic scheduling of a micro-grid with renewable energy resources. J Cleaner Prod 87:216–226
7. Mirjalili S, Mirjalili SM, Lewis A (2014) Grey wolf optimizer. Adv Eng Soft 69:46–61
8. Karaboga D, Basturk B (2007) A powerful and efficient algorithm for numerical function optimization: artificial bee colony (abc) algorithm. J Global Optim 39(3):459–471
9. Storn R, Price K (1997) Differential evolution-a simple and efficient heuristic for global optimization over continuous spaces. J Global Optim 11(4):341–359
10. Dey B, Roy SK, Bhattacharya B (2019) Solving multi-objective economic emission dispatch of a renewable integrated microgrid using latest bio-inspired algorithms. Eng Sci Technol Int J 22:55–56
11. Khan NA, Awan AB, Mahmood A, Razzaq S, Zafar A, Sidhu GAS (2015) Combined emission economic dispatch of power system including solar photo voltaic generation. Energy Convers Manage 92:82–91

Chapter 7
Mixed-Integer Differential Evolution Algorithm for Optimal Static/Dynamic Scheduling of a Microgrid with Mixed Generation

Sunita Shukla and Manjaree Pandit

Abstract The objective of dispatching generating units in electrical power supply system is to compute an optimal generation schedule to minimize the cost without violating the operating limits. Earlier, this problem comprised mainly of fossil fuel generating units. Now, the system complexity increases due to the widespread involvement of large number of renewable distributed energy resources (DERs) which are random, uncertain and introduce discrete variables in the objective function. This chapter presents a static and dynamic optimal scheduling model for a microgrid comprising of diesel generators (DG), microturbine (Mt), wind turbine and solar photovoltaic (PV) plant. A mixed-integer differential evolution (MIDE) algorithm with continuous as well as binary variables is used to solve the optimal dispatch problem with various equality, inequality and binding dynamic constraints. The developed model is tested on a modified five DER microgrid, and its performance is validated by using binary PSO and existing results from the literature.

Keywords Mixed-integer differential evolution · Optimal scheduling · Wind–solar microgrid · Uncertainty costs

1 Introduction

Restructuring of the electrical power network has led to major changes in the operational policies of the power industry. The inception of microgrids (MG) has enabled the penetration of distributed energy resources (DERs) into the already existing grid powered by the conventional fossil fuel-based generating units. Microgrid also helps to enhance the interconnection of storage systems and flexible loads to effectively

S. Shukla (✉) · M. Pandit
Department of Electrical Engineering, Madhav Institute of Technology and Science, Gwalior, India
e-mail: sunshuks@gmail.com

M. Pandit
e-mail: manjaree_p@hotmail.com

M. Pandit et al. (eds.), *Nature Inspired Optimization for Electrical Power System*, Algorithms for Intelligent Systems,
https://doi.org/10.1007/978-981-15-4004-2_7

operate as a separate electrical utility or as an integral part of the grid. Inclusion of RES through MGs in the dispatch schedule can help mitigate the undesirable intermittent and uncertain behavioral characteristic resulting in a more resilient and reliable power system network.

Several optimization tools and methods have been developed to formulate the dispatch problem using microgrids. In [1], energy investment and management optimization were performed for an individual site MG comprising of fuel cell, reciprocating engine, microturbine with solar, thermal, PV and storage mix using DER-CAM (distributed energy resource customer adoption model) of Berkeley laboratory, USA. In [2], a study to improve technology and policy variables for MGs is made, and a DER-CAM cost optimization model is developed that emphasizes the use of optimal PV plant and storage units only as supplements to gas generators. In [3], the optimal scheduling and management of a sample 14-bus radial MG which has 4-DERs with utility grid-connected mode is tested and validated using DE algorithm. An islanded MG model with DG/Mt/WT/PV, battery and converters is simulated using a multicross learning-based differential evolution algorithm to optimally schedule the units in [4]. In [5], an environmental economic dispatch MG model is evaluated under the island, grid-connected and energy exchange modes using an attraction and repulsion-based modified imperialist competitive algorithm. Optimal scheduling of a PV/WT/energy storage coordinated (DG) MG model is solved using hybrid harmony search and DE algorithm with adaptive parameters, improved mutation and competition operators in [6], while a semi-Markov model is employed in [7] to coordinate PV units with fossil fuel units and energy storage. In [8], an improved artificial bee colony algorithm-based model for cost optimization and energy management of a microgrid problem has been proposed.

At present, due to the mixing of continuous and discrete variable-based DERs in the MG scheduling and dispatch problems, most MG systems are being configured as DC or hybrid AC/DC models and solved using real time [9, 10] or mixed integer/binary-based [11–13] optimization tools.

This chapter proposes a mixed-integer differential evolution algorithm (MIDE) to solve the optimal scheduling and dispatch problem for a wind–solar–diesel microgrid comprising of continuous and discrete variables. MIDE has been applied for two test cases, (i) single objective (SO) minimization in a static environment and (ii) two-objective optimization under dynamic conditions. The results are verified using binary PSO and [3].

2 Problem Formulation for Microgrid with Mixed Generation

The objective of the MG optimal scheduling problem is to optimize operating cost and other conflicting objectives such as emission, loss and wind penalty and reserve cost such that complex nonlinear continuous/discrete operational constraints are satisfied.

The cost minimization objective function can be formulated for a dynamic/static scheduling problem as follows taking N_g generators for a time T with k time intervals ($\sum k = 24$). Here, C_T is the total cost of generation; P_{Gi}^k is the power generated, and $C_i\left(P_{Gi}^k\right)$ is the cost function of the i-th generating unit calculated for k-th time interval where k is fixed under static conditions.

$$\text{Min}\, C_T = \sum_{i=1}^{N_g} \sum_{k=1}^{T} C_i\left(P_{Gi}^k\right) \tag{1}$$

Subject to the following constraints.

2.1 Generating Unit Limits

$$P_{Gi,\min}^k \leq P_{Gi}^k \leq P_{Gi,\max}^k \tag{2}$$

where $P_{Gi,\min}^k$ and $P_{Gi,\max}^k$ are the minimum and the maximum generation limits of the i-th unit.

2.2 Supply and Load Balance Constraint

$$\sum_{i=1}^{N_g} \left(P_{Gi}^k\right) = P_D^k + P_L^k \tag{3}$$

$$P_L^k = \sum_{i=1}^{N_g} \sum_{j=1}^{N_g} P_{Gi}^k B_{ij} P_{Gj}^k + \sum_{i=1}^{n} B_{0i} P_{Gi}^k + B_{00} \tag{4}$$

where P_D^k is the total load demand; P_L^k is the power system loss at the k-th time, and $B_{ij,} B_{0i}$, B_{00} are the power transmission loss coefficients.

2.3 Generator Ramp Rate Limits

$$P_{Gi}^k - P_{Gi}^{k-1} \leq R_{\text{up},i} \tag{5}$$

$$P_{Gi}^{k-1} - P_{Gi}^{k} \leq R_{\text{down},i} \tag{6}$$

where $R_{\text{up},i}$ and $R_{\text{down},i}$ are the maximum ramp up and down rates per hour for the i-th unit, respectively.

2.4 Formulation of Total Cost Function for the Wind–PV–Diesel Microgrid

The total cost function in (1) is the sum of fuel cost, wind power (WP) cost and the PV power costs. The fuel cost function, $\text{FC}_i\left(P_{Gi}^{k}\right)$, can be formulated as a quadratic equation using the cost coefficients, a_i, b_i, c_i.

$$\sum_{i=1}^{N_g}\sum_{k=1}^{T}\text{FC}_i\left(P_{Gi}^{k}\right) = \sum_{i=1}^{N_g}\sum_{k=1}^{T} a_i + b_i P_{Gi}^{k} + c_i\left(P_{Gi}^{k})^2\right)\frac{\$}{\text{h}} \tag{7}$$

Cost for generating wind power WC can be calculated as follows [14]:

$$\text{WC} = \text{directcost} + \text{penaltycost} + \text{reservecost} \tag{8}$$

where for the i-th WP unit the directcost $= d_i \times w_i$; where d_i is the direct cost coefficient; penaltycost accounts for the wind curtailment; reservecost is a wind forecast error penalty associated with overestimation of WP; v_u is the wind curtailment penalty coefficient, and v_{ov} is the cost coefficient due to overestimation of wind power. The penalty and reserve costs can be calculated as follows using the Weibul probability distribution function (pdf) $f_W(w)$ of WP random variable w.

$$\text{Penalty Cost} = v_u \int_{w_s}^{w_{r,}} (w - w_i) f_W(w)\text{d}w \quad \text{and}$$

$$\text{Reserve Cost} = v_{ov} \int_{w_r}^{w_s} (w_i - w) f_W(w)\text{d}w$$

Weibul pdf is used to portray the uncertain wind power random variable by transforming the wind speed random variable into continuous and discrete distributions based on regions defined by Eqs. (9)–(11). The detailed solution is available in [15].

$$w = 0 \,\text{for}\, v\langle v_{\text{in}} \,\text{and}\, v\rangle v_r \tag{9}$$

$$w = w_r \times \frac{v - v_{\text{in}}}{v_r - v_{\text{in}}} \,\text{for}\, v_{\text{in}} \leq v \leq v_r \tag{10}$$

$$w = w_r \text{ for } v_r \leq v \leq v_o \tag{11}$$

where w_r is the rated WP, v_{in}, v_r and v_o are the cut-in, rated and cut-out wind speeds, c and k_s are the Weibul scale and shape factors at a particular location; both are dimensionless.

PV unit generating cost PVcost for the i-th solar unit which is generating power $P^k_{gs,i}$ at k-th time instant can be computed using $U^k_{s,i}$ the binary variable representing on–off status, ambient and reference temperatures t_{amb} & t_{ref}, temperature coefficient of the solar panel t_{coeff}, incident solar radiation $S(k)$, per unit PVcost, PVrate and rated PV power P_r using (12) and (13) as

$$\sum_{i=1}^{N_s}\sum_{k=1}^{T} \text{PVcost}\left(P^k_{gs,i}\right) = P^k_{gs,i} \times U^k_{s,i} \times \text{PVrate}_i \tag{12}$$

$$P^k_{gs,i} = P_r \times (1 + (t_{\text{amb}} - t_{\text{ref}}) \times t_{\text{coeff}}) \times (S(k) \div 1000) \tag{13}$$

2.5 SO and Two-Objective Optimization Functions

The objective function for SO cost optimization can be achieved by substituting Eqs. (7), (8) and (12) in Eq. (1) as:

$$\sum_{i=1}^{N_g}\sum_{k=1}^{T} C_i\left(P^k_{Gi}\right) = \sum_{i=1}^{N_g}\sum_{k=1}^{T} \text{FC}_i\left(P^k_{Gi}\right) + \text{WC} + \sum_{i=1}^{N_s}\sum_{k=1}^{T} \text{PVcost}\left(P^k_{gs,i}\right) \tag{14}$$

For the simultaneous optimization of two objectives (cost and emission), for optimal scheduling of wind–solar–diesel MG, price penalty factor (PPF) approach is used to create an aggregated function. The PPF converts emission into cost and the two objectives into a SO one. Emission is expressed using coefficients, α_i, β_i, γ_i as:

$$\sum_{i=1}^{N_g}\sum_{k=1}^{T} E_i\left(P^k_{Gi}\right) = \alpha_i + \beta P^k_{Gi} + \gamma_i (P^k_{Gi})^2 \frac{\text{kg}}{\text{h}} \tag{15}$$

$$\text{Min}\left[\sum_{i=1}^{N_g}\sum_{k=1}^{T} C_i\left(P^k_{Gi}\right) + \text{PPF} \times \sum_{i=1}^{N_g}\sum_{k=1}^{T} E_i\left(P^k_{Gi}\right)\right] \tag{16}$$

3 Mixed-Integer Differential Evolution (MIDE)

Differential evolution (DE) method was proposed by Storn and Price [16] and is amongst the most powerful optimizing methods for continuous variables. However, original DE needs to be modified to accommodate discrete variables. In this chapter, a simple update mechanism is employed for representing PV output which is discrete in nature.

The problem dimension D consists of C number of continuous variables and B number of binary variables such that $D = C + B$. The decision variable is generated randomly between the corresponding lower/upper limits, which are 0–1 for the discrete variables. Out of the five mutation strategies available in classical DE, the one with the best statistical performance for the problem is selected after preliminary studies. For the present problem, the mutant vector Z was generated using DE/best/1 as follows:

$$Z_i(k+1) = x_{i,\text{best}}(k) + f_m\left[x_{i,r2}(k) - x_{i,r3}(k)\right] \tag{17}$$

In (17), Strategy II: DE/best/1 means global best solution, one difference term and binomial crossover; $i = 1,2,\ldots,$ NP is the population index of the individual, k is the iteration count; x is the target vector; r_1 and r_2 are mutually different integers, and f_m is the mutation factor varying between 0 and 2.

Crossover operator is used for increasing the population diversity of the perturbed parameter vector to help generate a trial vector, Y_{ij}, replacing certain parameters of the target vector. Index j represents the problem dimension D, CR is crossover rate, rand is a uniform random number in range [0 1].

$$Y_{ij}(k+1) = \begin{cases} Z_{ij}(k+1) & \text{if } (\text{rand}(j)) \le \text{CR or } (j = \text{rand int}(i)) \\ x_{ij(k)} & \text{if } (\text{rand}(j)) \le \text{CR or } (j \ne \text{rand int}(i)) \end{cases} \quad \text{where } \forall j = [1, 2 \ldots C] \tag{18}$$

$$Y_{ij}(k+1) = \begin{cases} 1 & \text{rand}_b > 0.5 \\ 0 & \text{otherwise} \end{cases} \quad \text{where } \forall j = [C+1, \ldots D] \text{ and } b \in B. \tag{19}$$

To ascertain inclusion of trial vector in the next generation population, a greedy selection is performed using the given criterion

$$x_i(k+1) = \begin{cases} Y_i(k+1) & \text{if } (f(Y_i(k+1))) < f(x_i(k)) \\ x_i(k) & \text{otherwise} \end{cases} \tag{20}$$

4 Results and Discussion

This section presents the solution of optimal wind–solar–diesel microgrid scheduling problem for SO as well as two-objective problem with continuous and discrete decision variables. The presence of PV generators introduces binary variables, and hence, an MIDE algorithm is proposed for solving this mixed-integer nonlinear optimization problem. The developed approach is tested for different operating conditions and validated using binary PSO and [3]. Simulations were carried out using MATLAB 2013 on a1.6 GHz, core i3 processor with 8 GB RAM.

4.1 Description of the Modified Microgrid Test System

The microgrid with four DERs and utility [3] is modified by replacing utility grid by an equivalent DG, one DG by a wind turbine and one microturbine by a PV plant. The modified data of the 5-DER test system is given in Tables 1, 2, 3 and 4. B-coefficients in Eq. (4) are also taken from [3]. The solar data are taken from [17].

Table 1 Cost, emission coefficients and capacity limits for DERs

Bus no.	1	2	6	11
DER capacity (kW)	500 (DG1)	200 (WT)	80 (Mt)	100 (DG2)
a_i	10.193	–	0.5768	1.18
b_i	105.18	60.28	57.783	65.34
c_i	62.56	–	133.09	44.0
α_i	26.55	–	3.0358	19.38
β_i	16.1836	–	57.3403	176.6946
γ_i	7.0508	–	311.5728	821.6573
P_{gi}max	500	200	80	100
P_{gi}min	0.00	0.00	16	20
UR	200	70	30	40
DR	150	70	30	40
Heat rate (kJ/kWh)	10,314	0	11,373	10,581

Table 2 Wind speed data for DER-2

Rated capacity (kW)	Cut-in speed (miles/h)	Cut-out speed (miles/h)	Rated speed (miles/h)
200	5	25	15

Table 3 Data for the PV plant (DER-5)

Rated capacity (kW)	PVrate ($/h)	T_{ref} (°C)	T_{coeff} (% °C)
30	0.24	25	−0.005

Table 4 Solar radiation for 24-h period for PV plant (DER-5)

Hour	1	2	3	4	5	6	7	8
S (W/m^2)	0	0	0	0	5.4	101	253.7	541.2
T_{amb} (°C)	30	29	28	28	28	29	29	31
Hour	9	10	11	12	13	14	15	16
S (W/m^2)	530.4	793.9	1078	1125.6	1013	848.2	726.7	654
T_{amb} (°C)	33	34	35	36	37	37	37	38
Hour	17	18	19	20	21	22	23	24
S (W/m^2)	38	37	35	34	34	33	32	32
T_{amb} (°C)	392.9	215.1	38.5	0	0	0	0	0

4.2 Setting of the Optimal Parameters of MIDE

To select the optimal combination of population size (NP) and number of iterations (NI), tests were conducted for a load of 350 kW by varying population from 50 to 250 and the iteration count from 100 to 300. The performance is compared in Table 5 using statistical metrics computed out of 25 trial runs for each combination of NP and NI. It can be seen from Table 5 that as NP and NI are increased, and the consistency of the algorithm improves with reductions in mean, maximum and standard deviation (SD) values. However, the computational time is observed to increase in an obvious manner. Though the SD is observed to improve continuously with the increment in NP as well as NI, but in order to strike a balance between time and consistency, NP = 150 and NI = 300 (shown bold) are selected as optimal settings for conducting all further studies in this chapter. Increasing the two values beyond the tested settings resulted in unduly large CPU times with no significant improvement in standard deviation values. The mutation and crossover rates are both set at 0.5. The consistency of MIDE for cost optimization is compared in Fig. 1. The costs are quite consistent for all values of NP > 50. The convergence characteristic for different population sizes is shown in Fig. 2 which also shows a stable convergence of the MIDE algorithm for all the settings.

4.3 SO Optimal Static Scheduling of Microgrid Using MIDE

The static SO optimal schedule of all DERs has been computed for different loading conditions and tabulated in Table 6. DER-5 is a PV plant which has a binary status and

Table 5 Performance of MIDE with changing population and iteration count for load 350 kW

NP	Iteration (NI)	Mean cost	Maximum cost	St. Deviation	CPU time
50	100	130.00	626.2781	76.12	107.94
	150	63.38	243.62	15.69	158.80
	200	56.30	78.93	4.87	211.32
	250	57.02	74.23	4.90	267.33
	300	56.29	70.14	5.41	316.88
100	100	68.56	176.70	19.29	215.10
	150	59.32	101.95	8.28	334.56
	200	53.32	74.12	4.06	428.45
	250	50.89	67.14	2.06	536.91
	300	52.00	59.38	2.25	627.49
150	100	55.98	79.51	5.94	308.69
	150	55.82	69.63	3.88	527.2
	200	51.91	62.71	2.01	674.24
	250	51.90	58.77	2.27	845.91
	300	**50.99**	**58.07**	**1.73**	**1029.38**
200	100	54.11	60.89	3.05	474.17
	150	52.28	62.12	2.89	634.7
	200	50.90	54.70	1.26	917.59
	250	50.18	57.20	2.29	1114.56
	300	50.66	60.22	2.10	1341.23
250	100	54.40	64.49	3.22	669.77
	150	51.75	61.44	2.52	872.50
	200	51.00	61.30	2.06	1438.16
	250	49.88	54.34	1.56	1810.72
	300	49.44	52.55	1.58	1979.43

contributes its full capacity as given by Eq. (13). The proposed MIDE algorithm has computed the optimal DER schedule for minimum operating cost for eight different intervals as shown in Table 6. The nonlinear inequality and equality constraints given by Eqs. (2) and (3) are satisfied for all load cases of the mixed-integer nonlinear optimization problem. The corresponding values of all objectives, fuel cost, emission, wind cost, PVcost and loss are also computed for the optimal cost solution. Due to space limitations, similar results for SO optimization of each of these other objectives using MIDE are not given here.

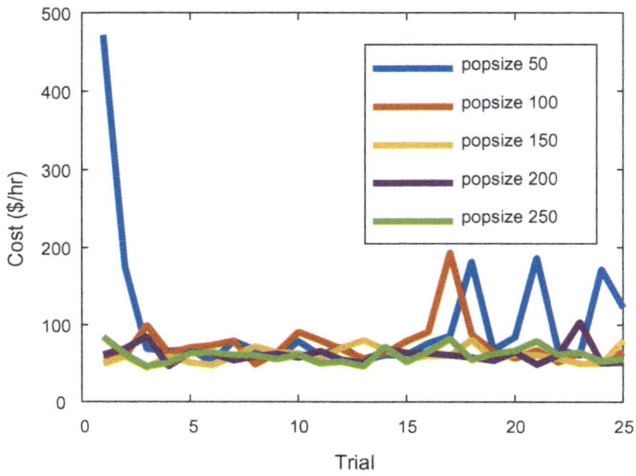

Fig. 1 Consistency of MIDE algorithm with variation in NP

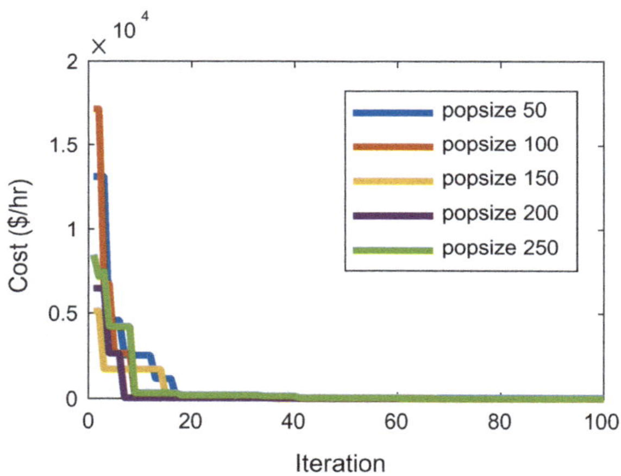

Fig. 2 Comparison of convergence characteristic of MIDE for different population sizes

4.4 SO Optimal Dynamic Scheduling of Wind–PV–Diesel Microgrid

The SO dynamic economic dispatch model was tested for a time horizon of 24 h for optimizing the cost given by Eq. (14). Hourly variation of the schedules and costs of the various DERs are shown in Figs. 3 and 4, respectively. From Fig. 3, it can be deduced that DER-2, i.e., WP contributes more than other DERs throughout the day. Participation of the microturbine and DG2 is almost constant over the day; however, DG1 share fluctuates throughout the day. Due to dependence on solar radiation, the PV contribution is limited to a few hours only. Figure 4 shows the variation of

Table 6 Microgrid static optimal scheduling for single objective cost function

Load (kW)	DG1 (kW)	WP (kW)	Mt (kW)	DG2 (kW)	PV (kW)	Us	PVcost	WC ($/h)	FC ($/h)	EmC (kg/h)	Loss (kW)	Tcost ($/h)
297	0.00	151.78	80.00	70.95	0.16	1	0.04	9.97	29.73	37.97	5.88	39.74
302	0.00	148.06	80.00	72.19	7.46	1	1.79	9.82	29.59	37.90	5.71	41.20
310	0.00	190.17	57.33	71.70	0.00	0	0.00	11.73	31.2	38.26	5.20	42.94
350	52.40	99.99	80.00	97.71	22.75	1	6.14	8.27	34.24	36.12	4.84	48.65
430	93.87	198.37	66.06	73.76	3.07	1	0.74	12.16	42.63	36.52	5.13	55.53
460	131.5	163.51	80.00	67.40	21.64	1	7.66	10.04	44.50	36.19	4.42	59.56
465	171.4	112.29	76.98	82.18	28.58	1	6.86	8.60	47.92	34.86	6.51	63.38

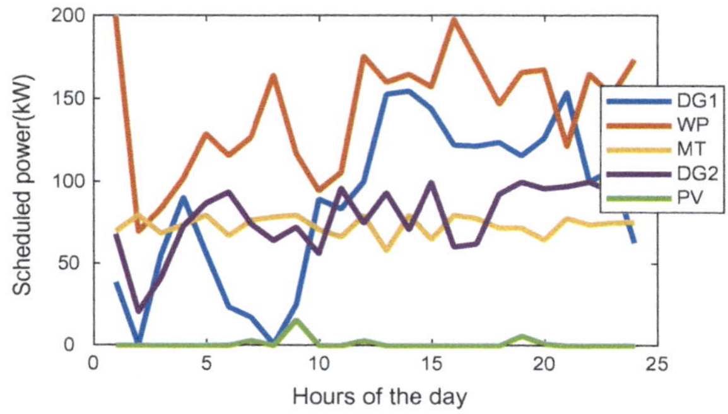

Fig. 3 Hourly variation of the optimal DER schedules

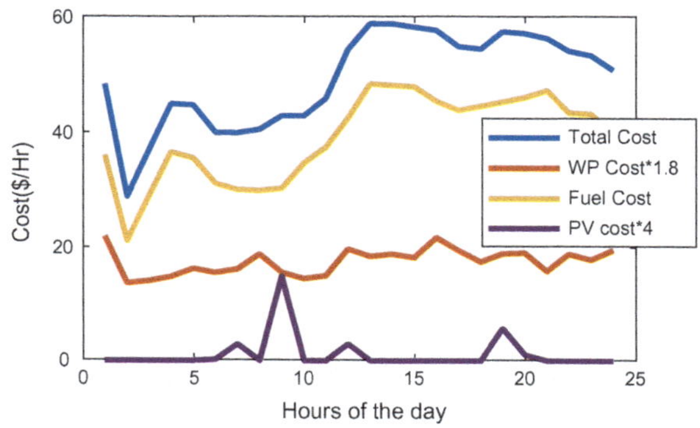

Fig. 4 Hourly share of DER costs

operating cost of DERs. Though the total capacity of renewable generation by PV and wind units is small as compared to the fuel-based DRs and Mt, their effect in reducing the total cost can be clearly observed. In the dynamic dispatch problem, additional constraints given by Eqs. (5) and (6) are imposed on the DER schedules.

4.5 Two-Objective Dynamic Optimal Scheduling of Wind–PV–Diesel Microgrid

After testing the performance of the MIDE for SO static scheduling problem, the more complicated two-objective case given by Eq. (16) was solved for dynamic operating conditions, which are closer to the practical MG operation. The optimal schedule of each hour is constrained by the previous optimal schedule as specified by ramp rate limits given by Eqs. (5) and (6). These additional dynamic constraints add further complexity to the MG scheduling problem. However, the proposed MIDE was found to produce a stable convergence for the tested cases. Table 7 gives the complete solution obtained by MIDE algorithm for the two-objective dynamic dispatch problem.

Table 7 Microgrid dynamic optimal schedule for combined cost and emission optimization

Load (kW)	DG1 (kW)	WP (kW)	Mt (kW)	DG2 (kW)	PV (kW)	Us	PVcost ($/h)	WC ($/h)	FC ($/h)	Em (kg/h)	Loss (kW)	Total cost (kg)
169	31.1	4.5	47.0	89.1	0.0	0	0.0	7.7	24.2	37.2	2.8	31.8
248	59.8	16.0	77.0	98.0	0.0	0	0.0	7.6	29.9	36.0	2.8	37.5
338	109.0	62.2	79.6	90.3	0.0	0	0.0	7.6	37.9	35.4	3.2	45.5
350	150.6	37.6	79.7	88.8	0.0	0	0.0	7.5	41.4	34.9	6.7	48.9
297	123.1	7.7	74.8	97.1	0.0	0	0.0	7.6	36.7	35.1	5.8	44.3
295	81.2	48.7	70.2	97.5	0.0	0	0.0	7.5	34.0	35.8	2.6	41.5
302	90.8	37.0	77.5	100.	0.0	0	0.0	7.5	34.9	35.5	3.2	42.4
308	64.9	76	75.3	98.6	0.0	0	0.0	7.7	33.8	36.0	2.4	41.5
310	88.8	66.0	75.4	81.9	0.0	0	0.0	7.6	35.0	36.1	2.1	42.7
350	93.0	63.4	73.8	100.0	22.7	1	5.5	7.6	36.6	35.5	2.9	49.7
430	198.0	70.8	80.0	88.5	3.1	1	0.7	7.7	49.4	34.2	10.3	57.8
460	210.6	110.9	62.8	85.5	0.0	0	0.0	8.6	52.6	34.4	9.9	61.1
465	207.2	104.5	68.1	94.8	0.0	0	0.0	8.4	52.6	34.1	9.7	61.0
462	151.2	124.9	80.0	86.3	23.9	1	4.9	9.0	46.6	35.0	4.9	61.3
455	213.6	93.9	71.0	87.6	0.0	0	0.0	8.1	52.4	34.2	11.1	60.6
430	188.3	87.8	74.3	88.0	0.0	0	0.0	8.0	48.9	34.4	8.5	56.9
432	178.1	78.5	71.7	100.0	11.0	1	2.6	7.8	47.8	34.3	7.5	58.3
456	227.5	64.8	76.0	95.6	6.1	1	1.5	7.5	54.0	33.6	16.3	63.0
452	149.4	126.8	80.0	99.6	1.1	1	0.3	9.0	47.4	34.7	4.8	56.7
446	175	98.0	80.0	100.0	0.0	0	0.0	8.2	48.9	34.3	6.9	57.1
435	181.7	92.4	75.5	93.1	0.0	0	0.0	8.1	48.8	34.4	7.6	56.9

(continued)

Table 7 (continued)

Load (kW)	DG1 (kW)	WP (kW)	Mt (kW)	DG2 (kW)	PV (kW)	Us	PVcost ($/h)	WC ($/h)	FC ($/h)	Em (kg/h)	Loss (kW)	Total cost (kg)
426	203.1	54.4	80.0	100.	0.0	0	0.0	7.5	49.9	33.9	11.5	57.5
407	182.8	70.2	69.7	92.5	0.0	0	0.0	7.7	47.3	34.4	8.2	55.0
370	181.1	53.7	45.0	98.3	0.0	0	0.0	7.5	45.5	34.9	8.1	53.0
Emission/day			Total WC/day		Total PVcost/day				Total FC/day			Total cost
838.3 kg			$189		$15.5				$1037			$1242

The solution satisfies the ramp up/down limits along with other continuous/discrete constraints.

5 Comparison and Validation of Results

In Table 8, the cost optimization results of MIDE (NP = 150, NI = 300) are compared with DE [3] (NP = 60, NI = 1500) for three loading conditions. MIDE algorithm is seen to produce comparable results with the DE of [3] with lesser number of function evaluations. In Table 9, optimal dynamic schedule and dispatch results of proposed MIDE algorithm have been compared with Binary PSO (BPSO) for three loading conditions of the modified microgrid system. In BPSO, NP = 5000, NI = 100. The results obtained by MIDE algorithm are found to be comparable to both classical DE and BPSO while fully satisfying all the constraints.

Table 8 Comparison of MIDE with conventional DE

Method	Load (kW)	Fuel cost ($/h)	DG1 (kW)	DG2 (kW)	Mt1 (kW)	DG3 (kW)	Mt2 (kW)
DE	169	23.97	0.00	63.20	80.00	20.30	6.30
	248	29.38	0.00	110.10	80.00	29.10	30.00
	338	35.90	0.00	166.30	80.00	64.30	30.00
MIDE	169	23.98	0.00	61.63	79.43	22.85	6.00
	248	29.36	0.00	94.77	79.21	46.71	29.27
	338	35.96	0.00	150.26	80.00	86.33	28.24

Parameters: MIDE: NP = 150, NI = 300 and DE: NP = 60, NI = 1500

Table 9 Comparison of the BPSO and MIBDE cost objective functions for the modified MG

Method	Load (kW)	Fuel cost ($/h)	DG1 (kW)	WP (kW)	Mt (kW)	DG2 (kW)	PV (kW)
BPSO	308	29.82	94.76	69.14	50.66	79.82	15.75
	350	34.71	110.76	80.72	49.66	88.86	22.75
	460	46.20	138.96	188.88	26.29	88.89	21.64
MIBDE	308	29.34	1.55	180.38	77.19	39.96	15.75
	350	34.24	52.40	99.99	80.00	97.71	22.75
	460	44.50	131.59	163.51	80.00	67.40	21.64

Parameter description: NP = 5000; D = 5; alpha = 10,000,000; NI = 100; trial = 25

6 Conclusion

The optimal scheduling of a wind–PV–diesel microgrid is presented for static/dynamic operation as well as for single/two-objective formulation. Due to the binary status of the solar PV output, the wind–PV–diesel MG scheduling and dispatch problem becomes a mixed-integer nonlinear problem comprising of continuous as well as discrete variables. Hence, a mixed-integer differential evolution algorithm is proposed for solving the problem that has complexities in the form of probabilistic and discrete terms in the objective function. The MIDE algorithm has a very simple formulation, and it is found to produce stable convergence, consistent and optimal results while satisfying complex nonlinear constraints. The results of MIDE are validated and found to be comparable with published results. The performance of MIDE is observed to be superior to the binary version of the PSO.

References

1. Marnay C, Venkataramanan G, Stadler M, Siddiqui A, Firestone R, Chandran B (2008) Optimal technology selection and operation of microgrids in commercial buildings. IEEE Trans Power Syst 23:975–982
2. Hanna R, Ghonima M, Kleissl J, Tynan G, Victor DG (2017) Evaluating business models for microgrids: Interactions of technology and policy. Energy Policy 103:47–61
3. Basu AK, Bhattacharya A, Chowdhury S, Chowdhury SP (2012) Planned scheduling for economic power sharing in a CHP-based micro-grid. IEEE Trans Power Syst 27:30–38
4. Hemmati M, Amjady N, Ehsan M (2014) System modeling and optimization for islanded micro-grid using multi-cross learning-based chaotic differential evolution algorithm. Int J Electr Power Energy Syst 56:349–360
5. Rabiee A, Sadeghi M, Aghaei J (2018) Modified imperialist competitive algorithm for environmental constrained energy management of microgrids. J Clean Prod 202:273–292
6. Zhang J, Wu Y, Guo Y, Wang B, Wang H, Liu H (2016) A hybrid harmony search algorithm with differential evolution for day-ahead scheduling problem of a microgrid with consideration of power flow constraints. Appl Energy 183:791–804
7. Barnes AK, Balda JC, Escobar-Mejia A (2015) A semi-markov model for control of energy storage in utility grids and microgrids with PV generation. IEEE Trans Sustain Energy 6:546–556
8. Roy K, Krishna K, Chandra A (2016) Modeling and managing of micro grid connected system using improved artificial bee colony algorithm. Int J Electr Power Energy Syst 75:50–58
9. Elsied M, Oukaour A, Youssef T, Gualous H, Mohammed O (2016) An advanced real time energy management system for microgrids. Energy 114:742–752
10. Li P, Han P, He S, Wang X (2017) Double-uncertainty optimal operation of hybrid AC/DC microgrids with high proportion of intermittent energy sources. J Mod Power Syst Clean Energy 5:838–849
11. Lúcia HLL, Leandro VRA, Julio M (2018) Optimization of Grid-Tied microgrids under binomial differentiated tariff and net metering policies: a Brazilian case study. J Control Autom Electr Syst. https://doi.org/10.1007/s40313-018-0403-x
12. Gil-gonzález W, Danilo O, Holguín E, Garces A, Grisales-noreña LF (2019) Economic dispatch of energy storage systems in dc microgrids employing a semidefinite programming model. J Energy Storage 21:1–8

13. Heris FN, Mohammadi B, Nazarpour D (2019) Economic dispatch of renewable energy and CHP—based multi-zone microgrids under limitations of electrical network. Iran J Sci Technol Trans Electr Eng. https://doi.org/10.1007/s40998-019-00208-4
14. Hetzer J, Yu DC, Bhattarai K (2008) An economic dispatch model incorporating wind power. IEEE Trans Energy Convers 23:603–611
15. Mohan H, Pandit M, Panigrahi BK (2015) Hybrid flower pollination algorithm with time-varying fuzzy selection mechanism for wind integrated multi-objective dynamic economic dispatch. Renew Energy 83:188–202
16. Storn R, Price K (1995) Differential evolution—a simple and efficient adaptive scheme for global optimization over continuous spaces. International Computer Science Institute, Berkeley, CA, Technical Report TR-95–012
17. Khan NA, Awan AB, Mahmood A, Razzaq S, Zafar A, Sidhu GAS (2015) Combined emission economic dispatch of power system including solar photo voltaic generation. Energy Convers Manag 92:82–91

Chapter 8
NSGA-II Based Reactive Power Management in Radial Distribution System Integrated with DGs

Himmat Singh and Laxmi Srivastava

Abstract This chapter presents Non-dominated Sorting Genetic Algorithm-II (NSGA-II), a swarm intelligence-based optimization technique, to solve Multi-Objective Reactive Power Management (MORPM) problem for minimization of active power losses, improvement of voltage profile, and minimization of total capacity of Reactive Power Sources (RPS) in Radial Distribution Systems (RDS). In an RDS having Distributed Generators (DGs), reactive power management problem can be solved by regulating reactive powers of DGs and of the reactive power compensation devices like FACTS devices, capacitor banks, etc. Efficacy of the proposed NSGA-II has been established by solving MORPM problem in IEEE 33-bus RDS penetrated with DGs and RPS units and by comparing the multi-objective reactive power management results with those obtained using multi-objective dragonfly algorithm, multi-objective differential evolution algorithm, and modified differential evolution algorithm.

Keywords Active power loss minimization · Non-dominated Sorting Genetic Algorithm-II · Total capacity of reactive power source units

1 Introduction

Integration of distributed generators (DGs) using has become essential in the present-day power systems to meet out the ongoing rapid social/economic growths and environmental challenges and to assure higher reliability of service and quality power. Hence, penetration of distributed generators in power system is increasing day-by-day. Distributed generators are located close to the load and hence are capable of

H. Singh (✉) · L. Srivastava
Department of Electrical Engineering, Madhav Institute of Technology and Science, Gwalior, India
e-mail: ahirwar.himmat@gmail.com

L. Srivastava
e-mail: srivastaval@hotmail.com

M. Pandit et al. (eds.), *Nature Inspired Optimization for Electrical Power System*, Algorithms for Intelligent Systems,
https://doi.org/10.1007/978-981-15-4004-2_8

delivering energy with high efficiency. Distributed generation consists of the generators based on renewable energy sources like wind turbines, biomass, solar energy, mini- and micro-hydroelectric plants, combustion turbines, fuel cells, etc. The use of alternative sources of energy reduces the dependency on scarce and perishable fossil fuels. Moreover, integration of distributed generation and consumption in a particular area increases the reliability and quality of power supply.

The DG technologies appear to be appealing, but it is an utmost important to analyze the impact of integrating them in a power network [1, 2]. To reduce active power loss, cost of transmission, and distribution of power and to improve system stability/ total voltage variations, etc., in a power system, distributed generators are placed on customer side or in distribution side of the network [3–6]. Various objectives of Reactive Power Management (RPM) can be achieved by optimizing the settings of generators' voltages/ reactive power, tap-settings of regulating transformers and settings of reactive power sources like FACTS devices, capacitor banks, etc [5, 6].

In a radial distribution system (RDS), DGs have significant effect on distribution of reactive power [5]. Hence, reactive power management or reactive power optimization in a RDS integrated with DGs is very important to make sure economic operation of the RDS without violation of operating limits and also to supply quality power to its customers.

As can be observed in literature review for reactive power management in the distribution system having DGs, a number of optimization methods have been proposed. Earlier, RPM problem was modeled as a single-objective constrained optimization problem (SORPM) and various Single-Objective Optimization (SOO) algorithms were employed to solve RPM problem [4]. Later on, two or more objective functions were handled simultaneously and RPM problem was formulated as a Multi-Objective Optimization (MOO) problem to achieve efficiently various objectives of the RPM problem [5–9].

Earlier, multi-objective RPM (MORPM) problem was transformed into a SOO problem using fuzzy approach weighted sum method [5, 7]. These methods simplify the MORPM problem significantly, but provide only one best solution for the MOO problem and do not provide other options to the decision maker [8]. Also, it does not reflect exactly the association among different constituent objectives.

Subsequently, some efficient multi-objective optimization techniques were employed to resolve MORPM problem handling multiple objectives simultaneously. Generally, the various objectives involved in MORPM are conflicting in nature, hence making it impossible to attain a solution that is able to optimize the involved objectives simultaneously. In place of providing one optimal solution, the MOO methods provide several non-dominated solutions [8–10].

This chapter proposes Non-dominated Sorting Genetic Algorithm-II (NSGA-II) for solving MORPM problem for radial distribution system (RDS) penetrated with DGs and reactive power source (RPS) units. The objective functions for the MORPM problem are minimization of active power loss, total capacity of RPS units, and total voltage variations, while decision or control variables are reactive power output of DGs and reactive power injected from the RPS units connected in the RDS.

Efficiency of NSGA-II is confirmed by solving MORPM problem using NSGA-II, multi-objective differential evolution (MODE) algorithm, multi-objective dragonfly (MOD) algorithm, and modified DE (MDE) algorithm in IEEE 33-bus RDS [11] having DGs and RPS units. The integrated DGs in the RDS are capable to deliver both the real and reactive powers.

2 Multi-Objective Reactive Power Management

Objectives of RPM in radial distribution systems having DGs are the active power loss minimization, total voltage variations minimization, and total capacity of the RPS units minimization, subject to various inequality and equality constraints [8, 9]:

2.1 *Objective Functions of RPM Problem*

Active Power Loss Minimization:
Active power loss P_{L} is the sum of active power loss occurring in different lines *nl* of the RDS and is expressed as

$$f_1 = P_L = \sum_{k=1}^{nl} G_k\left[V_i^2 + V_j^2 - 2V_iV_j\cos\theta_{ij}\right] \tag{1}$$

where G_k is conductance of any line *k*; $V_i\angle\theta_i$ and $V_j\angle\theta_j$ are the voltages at terminal buses *i* and *j* of *kth* line, respectively. $\theta_{ij} = \theta_i - \theta_j$.

Minimization of Total Voltage Variations:
Total voltage variations TV_V is minimized for improving the voltage profile of a distribution network and it can be accomplished by minimization of voltage magnitudes $|V_i|$ variations at various load (PQ) buses from reference voltage V_i^{ref} (1.0 pu). This objective function can be calculated as:

$$f_2 = TV_V = \sum_{i=1}^{\mathrm{LB}}\left|V_i - V_i^{\mathrm{ref}}\right| \tag{2}$$

where LB denotes the number of load buses in the distribution system.

Minimization of Total Capacity of RPS Units:
Total capacity of RPS units is minimized to reduce the expenditure on RPS units. Total capacity of the RPS units, $\mathrm{TC_{RPS}}$, is determined as

$$f_3 = \mathrm{TC_{RPS}} = \sum_{i=1}^{\mathrm{NQ}} Q_{\mathrm{RPS}i} \tag{3}$$

where $Q_{\mathrm{RPS}i}$ is the actual quantity of reactive power required from ith RPS unit and NQ is the total count of RPS units.

Equality and Inequality Constraints

Equality constraints considered in MORPM problem are power flow equations [21], expressed as:

$$\begin{cases} P_{gi} + P_{\mathrm{DG}i} - P_{li} = V_i \sum_{j=1}^{N_{\mathrm{bus}}} V_j \left(G_k \cos\theta_{ij} + B_k \sin\theta_{ij}\right) \\ Q_{gi} + Q_{\mathrm{DG}i} + Q_{\mathrm{RPS}i} - Q_{li} = V_i \sum_{j=1}^{N_{\mathrm{bus}}} V_j \left(G_k \sin\theta_{ij} + B_k \cos\theta_{ij}\right) \end{cases} \tag{4}$$

where P_g, P_{DG}, and P_l represent the active power output of the generators, the DG, and the load connected at bus i. Q_g, Q_{DG}, $Q_{\mathrm{RPS}i}$, and Q_l are reactive power output of the generators, the DG, the RPS unit, and the load connected at bus i. B_k stands for susceptance of line k in the distribution system having N_{bus} number of buses.

- Constraints for Buses and Feeder transmission capacity

To ascertain the quality power supply, voltage magnitude at any bus i is required to be inside the specified maximum and minimum limits. Mathematically,

$$V_i^{\min} \le V_i \le V_i^{\max}, \quad i = 1, 2, \ldots \mathrm{LB} \tag{5}$$

where $V_i^{\max}$ and $V_i^{\min}$ represent the maximum and minimum voltage limits, respectively. Also, the power flows through a distribution feeder should not exceed its specified thermal limit, expressed as:

$$S_k \le S_k^{\max}, \quad k = 1, 2, \ldots, nl \tag{6}$$

where S_k^{max} denotes the thermal limit of line k.

- Generators and RPS units Constraints

The constraints for the generators and the RPS units consist of the limits on the reactive power outputs of DGs and of the connected RPS units and can be written as [9]:

$$Q_{\mathrm{DG}i}^{\min} \le Q_{\mathrm{DG}i} \le Q_{\mathrm{DG}i}^{\min} \quad i = 1, 2, \ldots, \mathrm{NDG} \tag{7}$$

$$Q_{\mathrm{RPS}i}^{\min} \le Q_{\mathrm{RPS}i} \le Q_{\mathrm{RPS}i}^{\max} \quad i = 1, 2, \ldots, \mathrm{NQ} \tag{8}$$

Here, NDG stands for the quantity of RPS units. The MORPM problem may be expressed as (9), subject to various inequality constraints *h(x, u)* and equality constraints $g(x, u)$ and the decision variables' bounds as well

$$\text{Minimize } F = [f_1 f_2 f_3] \tag{9}$$

Subject to

$$h(x, u) \leq 0 \tag{10}$$

$$g(x, u) = 0 \tag{11}$$

and

Variable bounds

$$\text{ub}_j \geq u_j \geq \text{lb}_j \quad j = 1, 2, \ldots, n \tag{12}$$

Here, ub_j and lb_j are lower and upper bounds of *j*th control variable, respectively, and *n* is the number of decision variables. Various objectives of MORPM problem may be written as

$$f_1 = P_L(x, u), \quad f_2 = \text{TV}_V(x, u), \quad \text{and} \quad f_3 = \text{TC}_{\text{RPS}}(x, u)$$

Here, *x,* the dependent variables can be written as

$$x^{\text{T}} = [V_1 \ldots, V_{\text{LB}}, S_1, \ldots, S_{\text{nl}}] \tag{13}$$

and the control or independent variables *u* may be written as

$$u^{\text{T}} = \left[Q_{\text{DG1}}, Q_{\text{DG2}}, \ldots Q_{\text{DGNDG}}, Q_{\text{RPS1}}, Q_{\text{RPS2}}, \ldots Q_{\text{RPSNQ}}\right] \tag{14}$$

NSGA-II is applied for solving this combinatorial and nonlinear MORPM problem. Once the Pareto-optimal solutions are attained, the preferred solution is extracted out of these solutions using fuzzy membership function-based approach [12–14].

3 Non-dominated Sorting Genetic Algorithm-II for MORPM

Non-dominated Sorting Genetic Algorithm-II is one of the most efficient multi-objective optimization algorithms, proposed by Deb et al. [14, 15]. This algorithm is a fast and elitist version of non-dominated Sorting Genetic Algorithm, which is an evolutionary approach to find the optimal solutions for complex multi-objective optimization problems. Computational steps for NSGA-II to solve MORPM problem can be summarized in following steps:

i. Read the data of radial distribution system.
ii. Set maximum iterations (IT^{MAX}).
iii. Generate an initial population of NP, X_i (i = 1, 2,..., NP) inside control variables' upper and lower bounds.
iv. Set iteration IT =1.
v. For every individual, run Backward–Forward Sweep Power Flow (BFSPF) program [16] to calculate f_1, f_2, and f_3 using (1), (2), and (3).
vi. Calculate extended objective functions F_1, F_2, and F_3 using the objective functions f_1, f_2, f_3 and penalty factor (PF) as per constraints' violations as:

$$F_1 = f_1 + \text{PF}$$

$$F_2 = f_2 + \text{PF}$$

$$F_3 = f_3 + \text{PF}$$

vii. Sort out the population using non-dominated sorting.
viii. Apply selection, crossover, and mutation to obtain new solutions.
ix. The generation of size 2NP is formed by the parents and the off springs.
x. Merge the old set of solutions and newly created solutions to create population of size 2NP.
xi. A new generation of size NP is obtained by elitist sorting.
xii. If IT < IT^{MAX}, put IT = IT+1 and move to step v, otherwise move step xiii.
xiii. Stop. Find out the preferred solution from the Pareto-optimal solutions *NP* obtained in step xv.
xiv. Select the decision variables setting as per the preferred solution.

This decision variables setting will offer the minimum values of various objective functions for a given operating condition of the RDS.

4 Results and Discussion

To demonstrate the efficacy of NSGA-II, RPM problem has been solved in IEEE 33-bus systems integrated with DGs and RPS units optimally at different locations [11]. Performance of NSGA-II algorithm has been evaluated by applying NSGA-II [12–15], MODE [9, 6], MODA [17], and MDE algorithm [7] for the MORPM problem and comparing the results obtained for the three cases of both the radial distribution systems as:

Case 1: Minimization of P_L and TV_V
Case 2: Minimization of P_L and TC_{RPS}
Case 3: Minimization of P_L, TV_V and TC_{RPS}.

To validate the Pareto-fronts obtained using the developed NSGA-II and other EC/SI algorithms, a Reference Pareto-optimal front (POF) is generated by converting the MORPM problem to a SORPM problem and applying any SOO algorithm like GA, DE, or so on. In this chapter, modified DE algorithm [7] has been applied for attaining the Reference Pareto-front. The conversion of the MORPM problem into single-objective RPM problem has been done by the weighted combination of the normalized objective functions P_{LN}, TV_{VN}, and TC_{RPSN} as given below:

$$\text{Minimize}\, W \times P_{LN} + (1 - W) \times TV_{VN} \quad \text{in Case1} \tag{15}$$

$$\text{Minimize}\, W \times P_{LN} + (1 - W) \times TC_{RPSN} \quad \text{in Case2} \tag{16}$$

$$\text{Minimize}\, W_1 \times P_{LN} + W_2 \times TV_{VN} + W_3 \times TC_{RPSN} \quad \text{in Case3} \tag{17}$$

where W, W_1, W_2, and W_3 are weighing factors. The weighing factor W is *rand* [0,1], a random number distributed uniformly between 0 and 1. In (17), W_1, W_2, and W_3 are calculated by using 0.33, 0.33 and 0.34 times respectively the random number between 0 and 1.

MORPM problem is solved to find out the optimal settings of reactive power outputs of DGs and the shunt RPS units in the RDS. In all the three cases of the radial distribution system, the upper and lower voltage magnitude limits at all the load or PQ buses are 1.05 and 0.95 pu, respectively. In addition to the voltage constraints on load buses, line flow (feeder capacity) constraints are also imposed on MORPM problem. All the algorithms were implemented in MATLAB R2017a on Core i7 PC with 2.9 GHz, and 4 GB of RAM.

IEEE 33-bus RDS considered for MORPM problem is penetrated with three distributed generators of rating 794.8 kW, 1069 kW, and 1029 kW at bus nos. 13, 24 and 30, respectively [11]. In addition, this RDS contains three RPS at bus nos. 8, 18, and 30 [11] as depicted in Fig. 1. Active power loss values without DGs and without the RPS units are 210.98 and 72.83 kW, while total voltage variations values are 1.8044 pu and 0.6340 pu, respectively. Decision variables limits, i.e., lower and upper limits of the reactive power outputs of the DGs and RPS units are shown in Table 1.

With three DGs and three RPS units in IEEE 33-bus RDS, there are six decision variables. The proposed MOD algorithm has been applied to find out the best values of reactive power outputs of the three DGs and those of the reactive powers injected by the three RPS units integrated into the RDS for the three cases of MORPM. The best results achieved and provided here are for the population size *NP* equal to 40, number of iterations is equal to 500, crossover probability pc = 0.9, and mutation probability is pm = 0.17.

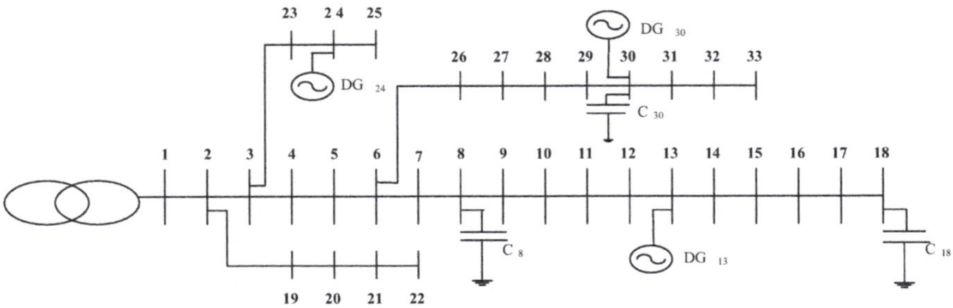

Fig. 1 Single-line diagram of IEEE 33-bus RDS [14]

Table 1 Decision variables limits

Parameter	Values (kVAR)
$Q_{\text{DG13}}^{\min}/Q_{\text{DG24}}^{\min}/Q_{\text{DG30}}^{\min}$	0.0
$Q_{\text{DG13}}^{\max}$	400
$Q_{\text{DG24}}^{\max}/Q_{\text{DG30}}^{\max}$	600
$Q_{\text{RPS8}}^{\min}/Q_{\text{RPS18}}^{\min}/Q_{\text{RPS30}}^{\min}$	0.0
$Q_{\text{RPS8}}^{\max}/Q_{\text{RPS18}}^{\max}$	450
$Q_{\text{RPS30}}^{\max}$	600

4.1 Case1: Minimization of P_{L} and TV_{V}

NSGA-II, MODE, MOD, and MDE algorithms have been employed for solving MORPM problem in IEEE 33-RDS with the objectives of minimizing P_{L} and TV_V. The POFs obtained using the MOO algorithms are compared with the reference POF obtained using MDE algorithm [7] in Fig. 2. This can be noted from Fig. 2, the POF

Fig. 2 Pareto optimal fronts for P_{L} and TV_V minimization (*Case* 1)

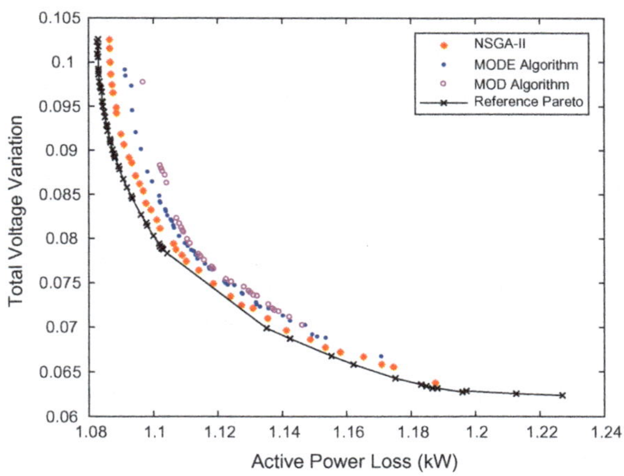

Table 2 Decision variable setting for preferred solution (Case 1)

Decision variables	Method			
	NSGA-II	MODE	MOD	MDE
Q_{DG13}	183.6601	196.8164	170.4849	163.6554
Q_{DG24}	503.0211	499.2734	480.0627	498.5274
Q_{DG30}	454.8102	403.1273	497.3783	501.2134
Q_{RPS8}	290.8213	287.6012	294.2970	299.9996
Q_{RPS18}	111.0904	101.7925	115.9758	118.1273
Q_{RPS30}	525.6614	571.8772	489.7839	479.6386
P_L	11.10199	11.0968	11.09589	11.07851
TV_V	0.077479	0.078552	0.078894	0.078323
TC_{RPS}	927.5731	961.2709	900.0567	897.7655

obtained using MOD algorithm is closest to the reference POF in comparison to those obtained using other MOO algorithms.

The preferred solutions as obtained using various algorithms along with TC_{RPS} are given in Table 2. From Table 2, this is clear that the developed NSGA-II gives the minimum P_L as 11.10,199 kW and minimum TV_V as 0.077479 pu which are close to those obtained in reference POF. It can be noted from Table II, that in case of MOD algorithm, the total capacity requirement of RPS units is the least. Thus, with the least total capacity of RPS units, the MOD algorithm provides minimum values of P_L and good voltage profile. This demonstrates the superiority of the developed MOD algorithm.

4.2 *Case 2: Minimization of* P_L *and* TC_{RPS}

In this case, various algorithms as mentioned in *Case1* were applied for minimization of P_L and TC_{RPS}. The reference POF and POFs achieved from various MOO algorithms are compared in Fig. 3. This can be noted in Fig. 3, the POF achieved using the proposed NSGA-II algorithm is closest to reference POF. The preferred solutions as obtained from various algorithms are compared in Table 3.

With the optimum setting of decision variables obtained in various MOO techniques, the TV_V was calculated and also given in Table 3. From Table 3, it is clearly observed that NSGA-II algorithm gives the minimum P_L as 11.5635 kW and minimum TC_{RPS} as 518.8817 kVAR which are close to those of Reference Pareto. Also, in case of MOD algorithm, the TV_V value is least.

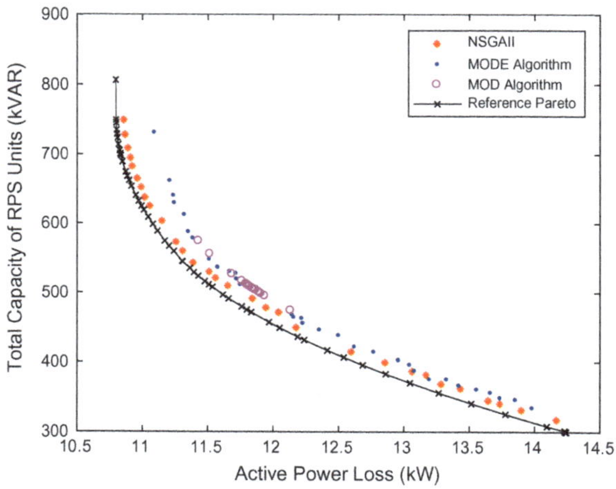

Fig. 3 Pareto optimal fronts for P_L and TC_{RPS} minimization (*Case 2*)

Table 3 Decision variable setting for preferred solution (Case 2)

Decision variables	Method			
	NSGA-II	MODE	MODA	MDE
Q_{DG13}	262.0402	266.2196	259.9947	259.9537
Q_{DG24}	519.0511	521.8735	480.2185	539.3594
Q_{DG30}	549.3313	546.9307	529.0573	550.0102
Q_{RPS8}	119.8303	103.3851	149.3667	100.3041
Q_{RPS18}	100.6313	100.2589	100.0100	100.0301
Q_{RPS30}	298.4201	308.6711	307.9197	315.7524
P_L	11.5635	11.5980	11.5626	11.5393
TC_{RPS}	518.8817	517.3151	519.4815	516.0866
TV_V	0.126146	0.128141	0.124384	0.128183

4.3 *Case 3: Minimization of* P_L*,* TV_V*, and* TC_{RPS}

Three objective functions considered in this case were minimization of P_L, TV_V, and TC_{RPS}. With these objective functions, NSGA-II, MODE, MOD, and MDE algorithms were applied to solve MORPM problem. The POFs attained using these algorithms are compared in Fig. 4. As can be seen from Fig. 4, the POF obtained using NSGA-II is very close to reference POF. The preferred solutions using various algorithms are compared in Table 4, which shows that the NSGA-II gives the minimum values of P_L as 11.37487 kW, TV_V as 0.087124 pu, and TC_{RPS} as 659.9941 kVAR, which are close to those of reference Pareto.

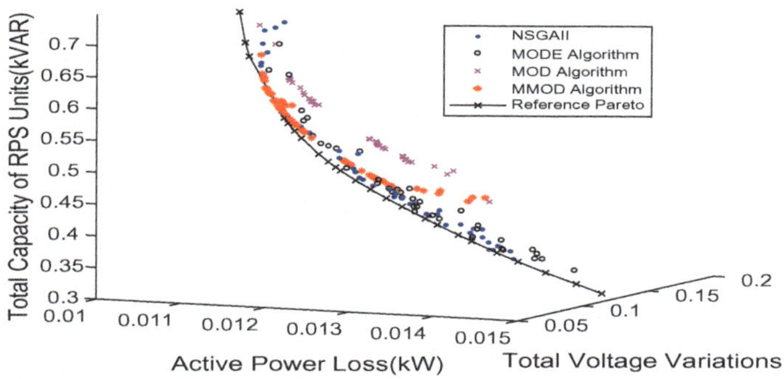

Fig. 4 Pareto optimal fronts for P_L, TV_V and TC_{RPS} minimization (*Case 3*)

Table 4 Decision variable setting for preferred solution (Case 3)

Decision variables	Method			
	NSGA-II	MODE	MODA	MDE
Q_{DG13}	290.8231	299.2393	279.2644	287.7976
Q_{DG24}	526.095	459.8671	364.7793	535.6955
Q_{DG30}	546.1467	497.9784	541.5244	549.9988
Q_{RPS8}	100.4428	152.1297	132.8907	100.0101
Q_{RPS18}	100.0421	103.3977	104.2023	100.0003
Q_{RPS30}	459.5092	383.5961	392.8250	358.7841
P_L	11.37487	11.50513	11.46121	11.38162
TV_V	0.087124	0.100523	0.103244	0.107213
TC_{RPS}	659.9941	639.1235	629.9180	558.7945

5 Conclusion

In this chapter, Non-dominated Sorting Genetic Algorithm-II has been proposed for solving MORPM problem in radial distribution system integrated with DGs and RPS units. For solving MORPM problem, the multiple objectives considered were active power loss minimization, voltage profile improvement, and minimization of the total capacity of RPS units. NSGA-II has been successfully employed for solving MORPM problem in IEEE 33-bus radial distribution systems penetrated with three DGs and three RPS units. Performance of NSGA-II is evaluated by comparing the obtained results with those offered by other MOO algorithms including MOD, MODE, and MDE algorithms. Comparison of the results clearly establishes the superiority of NSGA-II in terms of computational efficiency and solution quality and validates its potential for solving MORPM problem.

Acknowledgements The authors acknowledge financial support provided by TEQIP III. The authors also thank the Director M.I.T.S. Gwalior, India, for providing necessary facilities for carrying out this work.

References

1. Bollen MH, Hassan F (2011) Integration of distributed generation in the power system. In: IEEE Press Series on Power Engineering, vol 80, Wiley
2. Zhang L, Tang W, Liang J, Cong P, Cai Y (2016) Coordinated day-ahead reactive power dispatch in distribution network based on real power forecast errors. IEEE Trans Power Syst 31(3):2472–2480
3. Deshmukh S, Natarajan B, Pahwa A (2012) Voltage/VAR control in distribution networks via reactive power injection through distributed generators. IEEE Trans Smart Grid 3(3):1226–1234
4. Martinez-Rojas M, Sumper A, Gomis-Bellmunt O, Sudri-Andreu A (2011) Reactive power dispatch in wind farms using particle swarm optimization technique and feasible solutions search. Appl Energy 88:4678–4686
5. Niknam T, Narimani MR, Abarghooee RA (2013) Multiobjective optimal reactive power dispatch and voltage control A new opposition-based self-adaptive modified gravitational search algorithm. IEEE Syst J 7(4):742–753
6. Niknam T (2011) A new HBMO algorithm for multiobjective daily Volt/Var control in distribution systems considering distributed generators. Apply Energy 88:778–788
7. Singh H, Srivastava L (2014) Modified differential evolution algorithm for multi-objective VAR management. Int J Electr Power Energy Syst 55:731–740
8. Cheng S, Chen Y (2014) Multi-objective reactive power optimization strategy for distributionsystem with penetration of distributed generation. Int J Electr Power Energy Syst 55:731–740
9. Singh H, Srivastava L (2017) Multi-objective optimal reactive power dispatch for distribution system. In: 2017 IEEE international conference on ICPCSI–2017, Chennai, India, 21–22 Sept 2017
10. Injeti SK (2018) A Pareto optimal approach for allocation of distributed generators in radial distribution systems using improved differential search algorithm. J Electr Syst Inf Technol 5:908–927
11. Moradi MH, Zeinalzadeh A, Mohammadi Y, Abedini M (2014) An efficient hybrid method for solving the optimal sitting and sizing problem of DG and shunt capacitor banks simultaneously based on imperialist competitive algorithm and genetic algorithm. Electr Power Energy Syst 54:101–111
12. Li Y, Liao S, Liu G (2015) Thermo-economic multi-objective optimization for a solar-dish Brayton system using NSGA-II and decision making. Electr Power Energy Syst 64:167–175
13. Wang XD, Hirsch C, Kang S, Lacor C (2011) Multi-objective optimization of turbomachinery using improved NSGA-II and approximation model. Comput Meth Apply Mech Eng 200:883–895
14. Deb K, Pratap A, Agarwal S, Meyarivan T (2002) A fast and elitist multiobjective genetic algorithm NSGA-II. IEEE Trans Evol Comput 6:182–197
15. Deb K (2010) Multi-objective optimization using evolutionary algorithms. Wiley, Ltd
16. Augugliaro A, Dusonchet L, Favuzza S, Ippolito MG, Riva SE (2010) A backward sweep method for power flow solution indistribution networks. Int J Electr Power Energy Syst 32(4):271–280

17. Mirjalili S (2015) Dragonfly algorithm a new meta-heuristic optimization technique for solving single-objective, discrete, and multi-objective problems. Neural Comput Appl. https://doi.org/10.1007/s00521-015-1920-1
18. Roselyn JP, Devaraj D, Dash SS (2013) Multi-objective differential evolution for voltage security constrained optimal power flow in deregulated power systems. Int J Emerging Electr Power Syst. https://doi.org/10.1515/ijeeps-2013-0086, pp 1–17

Chapter 9
Short-Term Hydrothermal Scheduling Using Bio-inspired Computing: A Review

Khushboo Sharma, Hari Mohan Dubey, and Manjaree Pandit

Abstract Short-term hydrothermal scheduling (SHTS) problem comprises of scheduling together several hydro and thermal generation units such that objectives such as cost, emission, etc., can be optimized. Normally, the objective of SHTS is to minimize the fuel cost of the thermal units over a certain time of period while satisfying different operating constraints associated with thermal and hydro systems. Due to complex, nonlinear, multimodal and/or discontinuous nature of objective function, various bio-inspired optimization methods have been proposed to obtain the optimal dispatch solution for the hydrothermal systems of different dimensions and complexity levels. This chapter attempts to present a detailed review of the numerous bio-inspired optimization algorithms employed over the last two decades to solve the short-term SHT scheduling problem.

Keywords Short-term hydrothermal scheduling · Bio-inspired optimization · Operational constraints · Non-convex objective functions

Nomenclature

F_{mt}	Cost of generation of *m*th thermal plant at '*t*'
P^{s}_{mt}	Power generation of *m*th thermal unit at time '*t*'
a_m, b_m, c_m, e_m, f_m	Fuel cost coefficients of *m*th unit
N_s	Number of thermal power generation units

K. Sharma (✉) · H. M. Dubey · M. Pandit
Department of Electrical Engineering, Madhav Institute of Technology and Science, Gwalior, Madhya Pradesh, India
e-mail: Sharmakhushboo354@gmail.com

H. M. Dubey
e-mail: harimohandubeymits@gmail.com

M. Pandit
e-mail: drmanjareep@gmail.com

M. Pandit et al. (eds.), *Nature Inspired Optimization for Electrical Power System*, Algorithms for Intelligent Systems,
https://doi.org/10.1007/978-981-15-4004-2_9

$P_{\text{load}}(t)$	Total load demand at time '*t*'
P^s_{mt}, P^h_{nt}	Power generation of thermal and hydro units at time '*t*'
$P_{\text{loss}}(t)$	Total transmission losses of the system at time '*t*'
$C_{1n}, C_{2n}, C_{3n}, C_{4n}, C_{5n}, C_{6n}$	Coefficients of *n* hydro unit
$V_n t^h, Q_n t^h$	Volume and discharge of nth hydro unit at time '*t*'
$P^h_{(n,\min)}, P^h_{(n,\max)}$	Min. and max. power generation values of *n*th hydro unit
$P^s_{(m,\min)}, P^s_{(m,\max)}$	Min. and max. power generation values of *m*th thermal unit
$V^h_{(m,\min)}, V^h_{(m,\max)}$	Min. and max. values of reservoir volume of *m*th hydro plant
$Q^h_{(m,\min)}, Q^h_{(m,\max)}$	Min. and max. values of water discharge of *m*th hydro plant
V^h_{m0}, V^h_{m24}	Reservoir volume of *m*th hydro unit at time zero and twenty-four
$V^h_{(m,\min)}, V^h_{(m,\text{end})}$	Initial and final reservoir volume of *m*th hydro unit
HP	Hydro plants
TP	Thermal Plant

1 Introduction

The rising global demand of electrical power leads to the commissioning and construction of large number of power plants with different kinds of generating units. Their coordinated operation and scheduling poses a challenge before the power system operators due to complex characteristics and constraints. The SHTS problem mainly deals with the scheduling of hydro and thermal generating units to fulfill required power demand while satisfying a large number of nonlinear equality/inequality constraints. The operating cost of the hydro units is understandably almost negligible; hence, the primary aim of SHTS problem is minimization of fuel cost of thermal units in such a way that all operating constraints such as maintaining power and water balance, adherence to generation min-max limits, reservoir storage constraints, etc., are fully satisfied. Furthermore, valve-point loading (VPL) of thermal units and transmission losses makes the SHTS problem non-convex and nonlinear [1, 2]. The SHTS is seeking attention of the researchers from the last several years. Initially, various classical mathematical methods, namely nonlinear programming (NLP) [3, 4], linear programming (LP) [5], decomposition techniques (DT) [6, 7], dynamic programming (DP) [8] and Lagrange multiplier (LMP) [9, 10], have been introduced to obtain a better solution for the SHTS problems. However, conventional optimization methods which are usually analytical in nature are unable to solve complex problems. Hence, SHTS problem has been frequently solved using different heuristic and metaheuristic algorithms due to their flexible and versatile characteristics in dealing with non-convex, nonlinear, discontinuous and multimodal problems. Bio-inspired algorithms are mainly inspired by natural processes, and these may be classified into four categories. Evolution based, which employs operators such as reproduction, mutation, recombination and selection which are derived from evolution in nature. The examples of this class are genetic algorithm (GA) [11], differential evolution (DE) [12], evolutionary programming (EP) [13], etc. Second category can be algorithms which employ ecological processes in nature dealing with species and their distribution. Some of the algorithms based on ecology are cuckoo search algorithm (CSA) [14], flower pollination algorithm (FPA) [15]. The third class of algorithms is inspired by the physical processes existing or being followed in nature, for example gravitational search algorithm (GSA) [16] and lightening search algorithm. The fourth class is swarm intelligence-based algorithms, which are perhaps the most popular class and include a large number of algorithms which represents the collective behavior of social insects and animals such as particle swarm optimization (PSO) [17], artificial bee colony (ABC) [18]. A detailed classification of bio-inspired algorithms can be found in Ref [19]. Also, optimization methods applied for solving SHTS problem is given in [20]. This reference includes the brief description about heuristic and classical methods applied to solve SHTS problems.

In this chapter, the main idea is to present state-of-the art review of bio-inspired algorithms to solve a SHTS problem. This chapter discusses the formulation of SHTS problem with many different complexity constraints like ramp rate limit (RRL), VPL and prohibited discharge zone (PDZ), transmission losses. Also, the application of briefly described bio-inspired solution algorithms for SHTS problems of different dimensions is presented. The chapter organization is as follows: Sect. 2 gives the formulation of various objectives and constraints of the SHTS problem, and Sect. 3 presents an overview of various bio-inspired algorithms and compiles and tabulates the application of these algorithms for solving SHTS problems of different levels of complexities and dimensions.

2 Formulation of SHTS Problem

The idea behind the SHTS problem is to obtain hydro and thermal generator schedules such that the load demand is met while all operating limits and equality/inequality constraints are fulfilled while optimizing the fuel cost of thermal units.

2.1 Objective Function

The objective function given below by (1) [21–30] is a quadratic equation which represents the total fuel cost of thermal units at a given time ‘t’:

$$F_{mt}\left(P_{mt}^{s}\right) = a_m\left(P_{mt}^{s}\right)^2 + b_m P_{mt}^{s} + c_m \tag{1}$$

The actual fuel cost function is obtained when the VPL effect of thermal generating units is included by introducing a sinusoidal term in the function as given by [23–37]:

$$F_{mt}\left(P_{mt}^{s}\right) = a_m\left(P_{mt}^{s}\right)^2 + b_m P_{mt}^{s} + c_m + \left|e_m \sin\left(f_m\left(P_{m,\min}^{s} - P_{mt}^{s}\right)\right)\right| \tag{2}$$

For the dynamic scheduling, the fuel cost to be minimized is computed over a time period T consisting of small intervals of time 't' as [26, 27, 30, 38]:

$$F_{mt}\left(P_{mt}^{s}\right) = \sum_{t=1}^{24}\sum_{m=1}^{N_s}\left\{a_m\left(P_{mt}^{s}\right)^2 + b_m P_{mt}^{s} + c_m + \left|e_m \sin\left(f_m\left(P_{m,\min}^{s} - P_{mt}^{s}\right)\right)\right|\right\} \tag{3}$$

2.2 Operational Constraints

There are following constraints and limits of thermal and hydro generating units which must be satisfied while the objective functions are minimized.

Power Balance: The power balance equality constraint [21, 23–31, 33–36, 38–42] states that the total power generated by hydro and thermal plants must be equal to the demand and transmission losses at each time given as:

$$P_{\text{load}}(t) = \sum_{m=1}^{N_s} P_{mt}^{s} + \sum_{n=1}^{N_h} P_{nt}^{h} - P_{\text{loss}}(t) \tag{4}$$

The transmission losses in the system are expressed by using B-coefficients as [24–28, 34, 36]:

$$P_{\text{loss}}(t) = \sum_{u=1}^{N_s+N_h}\sum_{v=1}^{N_s+N_h} P_{ut} B_{uv} P_{vt} + \sum_{u=1}^{N_s+N_h} B_{0u} P_{ut}^{h} - B_{00} \tag{5}$$

The hydro power output is found by using a quadratic equation of reservoir volume and water discharge [21–46]:

$$P_{nt}^{h} = C_{1n}\left(V_{nt}^{h}\right)^2 + C_{2n}\left(Q_{nt}^{h}\right)^2 + C_{3n}\left(V_{nt}^{h} Q_{nt}^{h}\right) + C_{4n}\left(V_{nt}^{h}\right) + C_{5n}\left(Q_{nt}^{h}\right) + C_{6n} \tag{6}$$

Power Generation Limitations: The generation limits of hydro and thermal units are expressed by [21–38, 40–43, 46]:

$$P^h_{n,\min} \leq P^h_{nt} \leq P^h_{n,\max}; \quad n = 1, 2, \ldots, N_h$$

$$P^s_{m,\min} \leq P^s_{mt} \leq P^s_{m,\max}; \quad m = 1, 2, \ldots, N_s \tag{7}$$

Reservoir Storage Volume [21, 24–36, 38–46].

$$V^h_{n,\min} \leq V^h_{nt} \leq V^h_{n,\max}; n = 1, 2, \ldots, N_h \tag{8}$$

Water Discharge Limitation [23, 25, 28, 31, 34, 46]:

$$Q^h_{m,\min} \leq Q^h_{mt} \leq Q^{h(L,1)}_m$$

$$Q^{h(U,n-1)}_n \leq Q^h_{nt} \leq Q^{h(L,n)}_n; \quad n = 1, 2, \ldots, N_h$$

$$Q^{h(U,n)}_n \leq Q^h_{nt} \leq Q^h_{n,\max} \tag{9}$$

Water Balance [22, 23, 25–29, 31–36, 38, 39, 42]:

$$V^h_{nt} = V^h_{n(t-1)} + I^h_{nt} - Q^h_{nt} - S^h_{nt} + \sum_{n \in R^n_{up}} \left(Q^h_{n(t-\tau_n)} + S^h_{n(t-\tau_n)} \right) \tag{10}$$

The reservoir volume at initial and final level is represented as [22, 24–29, 31, 33, 35, 37, 39–42, 44–46]:

$$V^h_{n0} = V^h_{n,\min}$$

$$V^h_{n24} = V^h_{n,\text{end}} \tag{11}$$

3 Bio-Inspired Algorithm and Their Application

3.1 *Genetic Algorithm (GA)*

This is a population-based evolutionary algorithm which was proposed in 1989 [11] by Goldberg. This search technique employs operators like inheritance, mutation, selection, crossover, elitism to model the biological evolution process. This algorithm mainly deals with the genes which combine and make strings of chromosomes. The population of GA is generated by the chromosomes. The genes present in the chromosomes give all the required information related to the encoding which is done for the input variables of the problem depending on level of accuracy desired. The GA works on the principal of 'survival of the fittest.' The idea is to carry the members of the population which have a better fitness to the next generation while the ones with lower fitness are dropped from the mating pool.

To solve SHTS problems, GA uses the following steps: (i) initialization of population, (ii) compute the fitness function, (iii) selection operation, (iv) crossover and mutation. GA modeling framework and solution technique for SHTS is employed in [39]. GA is also used to solve the convex SHTS problems [21]. RCGA-AFSA [31] is a hybrid of real-coded GA and artificial fish swarm algorithm which combines the global search capability of RCGA and local search capability of AFSA to speed up convergence and improve the search ability (Table 1).

3.2 Particle Swarm Optimization (PSO)

The PSO is the most popular and widely used swarm intelligence-based algorithm proposed by Kennedy and Eberhart in 1995 [17] applied for solving complex practical problems across all engineering domains. PSO gets inspiration from the group foraging behavior of swarm of birds. In order to find best solution, particles fly randomly in search space. These particles then update positions following their own experience and neighbor's experience; in this manner, best solution can be found. PSO is a stochastic method, and it is very useful for the solution of continuous problems. It has been applied over a wide range of real-world problems. Implementation of PSO is simple, and it is insensitive in case of design variables scaling.

It has three selection parameters: (i) size of population (ii) constriction factor and (iii) acceleration factors.

PSO methods were employed for solution of SHTS problem in [22, 32] to provide near-optimal solutions with high efficiency and consistency. An improved PSO (IPSO) algorithm [43, 44] and modified adaptive PSO (MAPSO) [23] are also proposed for solving SHTS problem. IPSO makes convergence speed faster by reducing search space margin. The MAPSO [23] algorithm was found to handle discontinuous and non-convex problems efficiently.

The SHTS problem has also been solved by a modified version of dynamic neighborhood learning-based PSO (MDNLPSO) [24] which provided global search and exploration. Couple-based PSO (CPSO) for SHTS is employed in [25]; CPSO aims at overcoming the premature convergence problems.

3.3 Differential Evolution (DE)

The DE proposed by Storn and Price in 1997 [12] is the second most popular evolutionary algorithm which employs mutation of the initial randomly generated population to produce a mutant vector. There are five mutation strategies proposed which modify the target vector/best individual by adding vector difference of two random

Table 1 Dimension, complexity level of problem solved using various algorithms

Method	No. of hydro plants/units	No. of thermal units	Convex	RRL	PDZ	VPL	Loss
GA, PSO, DE, EP, ABC, GSA, CSA, TLBO, FPA	4 [31, 39, 22–25, 43, 44, 34, 35, 40, 41, 26, 27, 36, 37, 42, 29, 30, 38]	1 [31, 39, 22–25, 43, 44, 34, 35, 40, 41, 26, 27, 36, 37, 42, 29, 30, 38]	✓	✗	✗	✗	✗
GA, PSO, DE, EP, TLBO, FPA	4 [31, 23–25, 43, 34, 40, 41, 26, 27, 45, 29, 38]	1 [31, 23–25, 43, 34, 40, 41, 26, 27, 45, 29, 38]	✗	✗	✓	✓	✗
PSO, TLBO, FPA	4 [43, 29, 38]	1 [43, 29, 38]	✓	✗	✓	✗	✗
DE, TLBO	4 [35, 30]	1 [35, 30]	✗	✗	✗	✓	✗
DE	4 [34]	10 [34]	✗	✗	✗	✓	✗
GA	4 [21]	6 [21]	✓	✗	✗	✗	✗
GA, PSO, DE, ABC, GSA, TLBO	4 [31, 23–25, 32, 33, 41, 28, 37, 29]	3 [31, 23–25, 32, 33, 41, 28, 37, 29]	✗	✗	✗	✓	✗
GA, PSO, DE, EP	4 [31, 24, 25, 34, 40, 41, 26]	3 [31, 24, 25, 34, 40, 41, 26]	✗	✗	✗	✓	✓
GA, DE, ABC	4 [31, 34, 36]	3 [31, 34, 36]	✗	✓	✓	✓	✓

(continued)

Table 1 (continued)

Method	No. of hydro plants/units	No. of thermal units	Convex	RRL	PDZ	VPL	Loss
PSO	4 [23]	3 [23]	✗	✓	✓	✓	✗
EP	4 [27]	3 [27]	✗	✗	✓	✓	✓
ABC	4 [28]	3 [28]	✓	✗	✗	✗	✓
GSA	4 [46]	3 [46]	✓	✗	✗	✗	✗

individuals. The mutation factor controls the magnitude of mutation. After initialization and mutation, the population undergoes crossover of randomly chosen individuals based on a crossover factor to generate the trial vector. A greedy selection between target and trial vector is carried out to select the population pool for the next generation. In DE, the mutation takes place by adding weights to the vectors. It has three user-controlled parameters: (i) size of population, (ii) control parameter of crossover and(iii) amplification factor of difference vector.

DE carried out parameter study of hydrothermal system [33]; an improved DE for solving SHTS problems is introduced [34], to minimize the objective function. Also, to make DE more effective to solve SHTS problems with non-convex fuel function, it is hybridized with sequential quadratic programming (DE-SQP) [35]. In addition, a modified hybrid differential evolution with cascaded reservoirs (MHDE) is employed [40, 41], such that it does not need the penalty function and obtain optimal solution.

3.4 *Evolutionary Programming (EP)*

Evolutionary programming (EP) [13] is inspired by the evolution pattern of biological systems; many novel simulated evolutionary algorithms for the optimization have been developed rapidly in recent years. It has five selection parameters: (i) size of population (ii) rate of mutation (iii) rate of crossover (iv) size of tournament and (v) strategy parameter.

Evolutionary programming algorithm not only has a high convergence rate but also preserves the diversity of the population so as to get away from local optima. For the optimization of functions, evolutionary programming follows the following steps: (i) initialization, (ii) mutation operation, (iii) selection operation and (iv) determine whether termination condition is satisfied or not.

QEA is an evolutionary algorithm which is probabilistic and is based upon the quantum calculus so that the search becomes better and robust. Clonal real-coded QEA (CRQEA) having Cauchy mutation for SHTS is introduced in [26]. Real-coded rule in CRQEA is used for global optimization. In CRQEA, early convergence is avoided by clonal operator and Cauchy mutation. Therefore, CRQEA provides an effective approach to solve problem. SHTS using differential real-coded QEA (DRQEA) is introduced in [27].

A fast evolutionary programming technique for SHTS is introduced in [45]; it solves the problem with high convergence speed.

3.5 *Artificial Bee Colony (ABC) Algorithm*

The ABC is a swarm intelligence-based optimization technique proposed by Karaboga in 2005 [18] which is inspired by the way honeybees search for nectar. It has three selection parameters; (i) size of swarm (ii) number of onlooker, employed and scout bees and (iii) limit.

ABC is found to be more effective to solve complex engineering problems. Also, it has high convergence capability. There are three phases in this algorithm: the employed bee phase, where exploration is carried out, followed by the onlooker bee phase where the solutions found by employed bees are exploited, and the third phase is the scout bee phase in which new solutions replace the older ones which are not found to be stagnant for a predefined number of generations. Employed bees search the food sources, after that the employed bees circulate information about location of food sources with the onlookers, and finally the onlookers select the promising food sources by utilizing the information collected by employed bees.

In this manner, the better food sources move to the next generation through the employed/onlooker bee mechanism. A few employed bees are translated in to scouts that discard the available sources of food and then find out new sources of food. In this algorithm, there is two halves of the swarm in which employed bees come under first half and onlooker bees comes under second half. Employed bees are same in number as that of onlooker bees which further equals to number of food sources.

An ABC is employed in [36] to solve SHTS problem. Also, an adaptive chaotic ABC algorithm (ACABC) is employed in [28]. Moreover, chaotic searching behavior of bees is introduced for the avoidance of local optimum solution, whereas adaptive mechanism is used to avoid premature convergence of ABC in SHTS problem.

3.6 Gravitational Search Algorithm (GSA)

The GSA [16] is a bio-inspired technique which is based on law of gravity and mass interactions existing in nature. In GSA, natural objects represent optimization solutions and the mass of the objects represents the fitness of the solutions or their performance; i.e., the objects with heavy mass lead to good solutions and the objects with lower masses lead to the generation of poor solutions. According to the law of gravitational force, all the objects attract the other objects which lead to the global attraction of objects toward the object with heavier mass. It has four selection parameters: (i) size of population (ii) gravitational constant (iii) specified constant and (iv) threshold value constant. GSA performance is superior to solve nonlinear functions. GSA method uses the following steps to solve SHTS problems: (i) initialization, (ii) evaluation of fitness of agents, (iii) calculation of gravitational constant, (iv) update gravitational mass and inertial mass, (v) total force calculation (vi) velocity and acceleration calculation, (vii) update agents' positions, (viii) check solution's feasibility.

SHTS problem were solved using GSA in [46]. A hybridized GSA was introduced in [37] to solve SHTS problems, which includes certain equality, inequality constraints.

3.7 Cuckoo Search Algorithm (CSA)

The CSA is a bio-inspired optimization technique proposed by Yang and De in 2009 [14]. This algorithm is based on the parasitism behavior of the cuckoo birds which follow the practice (obligate brood parasitism) of putting their eggs in the nests of other birds for hatching. When used for solving optimization problem, each nest represents one solution. This algorithm has two user-defined tuning parameters: (i) size of population (ii) probability. The search process of CSA follows three well-defined rules which are: (i) at any one time, each cuckoo lays randomly selects one nest and lays one egg, (ii) the nest with better quality moves on to the next generation; the quality of the nest (solution) is defined by the quality of eggs and (iii) the cuckoo egg has a probability of being present in the host nests which are available. There are fixed number of host nests, and the host bird either builds a new nest or throws the alien cuckoo egg out of the nest to start a fresh.

The CSA has been proposed for solving the SHTS problem with cascaded hydropower plants [42] and has proved efficient by employing different probability distributions for generating new solutions.

3.8 Teaching-Learning-Based Optimization (TLBO)

The TLBO algorithm proposed by R. V. Rao et al. in 2011 [47] is a unique optimization technique based on the teaching–learning processes. It has no control parameters. The learners represent the population, and the algorithm works on how the influence and strength of teacher works to improve the quality of learners which is decided by their grades. There are two phases in this algorithm; the teacher phase in which the teachers transfers his knowledge to learners by sharing notes, etc., and the learner phase in which the best learner assumes the role of a teacher and is permitted to share his knowledge with other learners.

In TLBO, learners represent size of population, and distinct subjects are taken as design variables for optimization. Fitness of solution is reflected by the learner's results; fittest solution is identified as a teacher. Parameters which are used to solve the objective function are design variables, and best obtained solution is considered as the objective function's best value in the optimization problem.

The SHTS problem is attempted using TLBO in [29], and an effective methodology for SHTS using improved TLBO algorithm (ITLBO) is employed in [30]. The comparison of performance of TLBO and ITLBO has been carried out with other algorithms, and the effectiveness of TLBO-based methods has been depicted [30].

3.9 Flower Pollination Algorithm (FPA)

The FPA is a nature-inspired optimization algorithm which is based on the pollination process of flowering plants in nature. This algorithm is proposed by Xin-She-Yang in 2012 [15]. There are two control parameters: (i) size of population and (ii) switching probability. The reproduction in flowers is done by the pollination process through the transfer of pollen by the pollinators such as insects, birds, bats and other animals. The pollination process can be biotic or abiotic. Most of the pollination processes are biotic which involves pollination through insects and animals. The other kind of pollination is abiotic which does not need pollinator. Wind, diffusion and grass lead to the pollination process in this case.

Pollination can also be classified as cross-pollination and self-pollination. In case of cross-pollination, the pollens from the flowers of different plants are involved. These plants may be far away from each other; hence, this leads to the global exploration. In case of self-pollination, the pollens involved either come from the same flower or from the different flowers of same plant.

Improved FPA for SHTS (IFPA) is employed in [38]; in IFPA, a scaling factor is introduced to control the pollination process. In the improved FPA, the solution quality is improved by introducing an intensive exploitation phase. IFPA convergence characteristics are good when compared with other algorithms.

4 Conclusion

The chapter attempts to present a review of the bio-inspired algorithms used for solving the SHTS problem. The objective function of SHTS problems is non-convex, discontinuous and multimodal in nature. Moreover, the equality/inequality constraints involved are complex, with time dependant binding limits on variables.

Various bio-inspired optimization approaches have been applied for the solution of SHTS problem with different complexity levels. This chapter critically reviews different bio-inspired optimization, and their basic principle and advantage were discussed.

It is clearly observed that from table listed above, a large number of research work related to DE and PSO has been done. However, nowadays, other bio-inspired algorithms are also gaining popularity due to less number of control parameters and due to their global search capability.

In this review chapter, various bio-inspired algorithms are compared in various respects such as RRL, VPL, PDZ, transmission losses.

The chapter presents the formulation of the SHTS problems and classifies the reviewed literature on the basis of (i) the applied bio-inspired algorithm, (ii) level of complexity decided by the nature of objective function (convex/non-convex, continuous/discontinuous, with/without VPL, PDZ, etc., (iii) constraints mapped (inclusion of ramp rate limits, transmission losses, etc) and (iv) problem dimension given by the number of hydro and thermal power units. The review is expected to be useful to scholars and researchers interested in carrying out further research and investigations in this area.

Acknowledgements The financial support provided by AICTE-RPS project File No. 8-36/RIFD/RPS/POLICY-1/2016–17 dated 2.9.2017 and TEQIP III is sincerely acknowledged. Thanks are also due to the Director and management of MITS, Gwalior, India, for providing facilities and support.

References

1. Kumar S, Naresh R (2007) Efficient real coded genetic algorithm to solve the non-convex hydrothermal scheduling problem. Electr Power and Energy Syst 29:738–747
2. Gil E, Bustos J, Rudnick H (2003) Short-term hydrothermal generation scheduling model using a genetic algorithm. IEEE Trans Power Syst 18(4):1256–1264
3. Catalao JPS, Pousinho HMI, Mendes VMF (2011) Hydro energy systems management in Portugal: profit-based evaluation of a mixed-integer nonlinear approach. Energy 36:500–507
4. Catalao JPS, Mariano SJPS, Mendes VMF, Ferreira LAFM (2009) Scheduling of head-sensitive cascaded hydro systems: a nonlinear approach. IEEE Trans Power Syst 24(1):337–346
5. Chang GW, Aganagic M, Waight JG, Medina J, Burton T, Reeves S, Christoforidis M (2001) Experiences with mixed integer linear programming based approaches on short-term hydro scheduling. IEEE Trans Power Syst 16(4):743–749
6. Dos Santos TN, Diniz AL (2009) A new multi period stage definition for the multistage benders decomposition approach applied to hydrothermal scheduling. IEEE Trans Power Syst 24(3):1383–1392
7. Sifuentes WS, Vargas A (2007) Hydrothermal scheduling using benders decomposition: accelerating techniques. IEEE Trans Power Syst 22(3):1351–1359
8. Homem-de-Mello T, de Matos VL, Finardi EC (2011) Sampling strategies and stopping criteria for stochastic dual dynamic programming: a case study in long-term hydrothermal scheduling. Energy Systems 2(1):1–3
9. Dieu VN, Ongsakul W (2009) Improved merit order and augmented Lagrange Hopfield network for short-term hydrothermal scheduling. Energy Convers Manag 50:3015–3023
10. Yamin HY (2004) Review on methods of generation scheduling in electric power systems. Electr Power Syst Res 69:227–248
11. Goldberg DE (1989) Genetic algorithms in search, optimization, and machine learning. Addison-Wesley, Reading, MA
12. Storn R, Price K (1997) Differential evolution: a simple and efficient heuristic for global optimization over continuous spaces. J Glob Optim 11(4):341–359
13. Ya X, Liu Y, Lin G (1999) Evolutionary programming made faster. IEEE Trans Evolut Comput 3:82–102
14. Yang XS, Deb S (2009) Cuckoo search via lévy flights, in proc. World Congress on Nature & Biologically Inspired Computing 210–214
15. Yang XS (2012) Flower pollination algorithm for global optimization. In: Unconventional computation and natural computation 2012, Lecture Notes in Computer Science, vol 7445, pp 240–249
16. Rashedi E, Nezamabadi-pour H, Saryazdi S (2009) GSA: a gravitational search algorithm. Inf Sci 179(13):2232–2248
17. Kennedy J, Eberhart R (1995) Particle swarm optimization. In: Proceedings of IEEE international conference on neural networks, IV, pp 1942–1948

18. Karaboga D (2005) An idea based on honeybee swarm for numerical optimization (Tech.Rep. TR06), Erciyes University, Engineering Faculty, Computer Engineering Department
19. Dubey HM, Pandit M, Panigrahi BK (2018) An overview and comparative analysis of recent bio-inspired optimization techniques for wind integrated multi-objective power dispatch. Swarm Evol Comput 38:12–34
20. Labadie JW (2004) Optimal operation of multi reservoir systems: state-of-the-art review. https://doi.org/10.1061/(ASCE)0733-9496(2004)130:2(93)
21. Bisht VS (2012) Genetic algorithm solution for convex hydro-thermal generation scheduling problem. https://doi.org/10.1109/PowerI.2012.6479532
22. Yu B, Yuan X, Wang J (2007) Short-term hydro-thermal scheduling using particle swarm optimization method. Energy Convers Manag 48:1902–1908
23. Amjady N, Soleymanpour HR (2010) Daily hydrothermal generation scheduling by a new modified adaptive particle swarm optimization technique. Electr Power Syst Res 80:723–732
24. Rasoulzadeh-akhijahani A, Mohammadi-ivatloo B (2015) Short-term hydrothermal generation scheduling by modified dynamic neighborhood learning based particle swarm optimization. Electr Power Energy Syst 67:350–367
25. Wu Y, Wu Y, Liu X (2019) Couple-based particle swarm optimization for short-term hydrothermal scheduling. Appl Soft Comput J 74:440–450
26. Wang Y, Zhou J, Mo L, Yang S, Zhang Y (2012) A clonal real-coded quantum-inspired evolutionary algorithm with Cauchy mutation for short-term hydrothermal generation scheduling. Electr Power Energy Syst 43:1228–1240
27. Wang Y, Zhou J, Mo L, Zhang R, Zhang Y (2012) Short-term hydrothermal generation scheduling using differential real-coded quantum-inspired evolutionary algorithm. Energy 44:657–671
28. Liao X, Zhou J, Yang S, Zhang R, Zhang Y (2013) An adaptive chaotic artificial bee colony algorithm for short-term hydrothermal generation scheduling. Electr Power Energy Syst 53:34–42
29. Roy PK (2013) Teaching learning based optimization for short-term hydrothermal scheduling problem considering valve point effect and prohibited discharge constraint. Electr Power Energy Syst 53:10–19
30. Pasupulati B, Kumar RA, Asokan K (2017) An effective methodology for short-term generation scheduling of hydrothermal power system using improved TLBO algorithm. https://doi.org/10.1109/ICIEEIMT.2017.8116842
31. Fang N, Zhou J, Zhang R, Liu Y, Zhang Y (2014) A hybrid of real coded genetic algorithm and artificial fish swarm algorithm for short-term optimal hydrothermal scheduling. Electr Power Energy Syst 62:617–629
32. Mandal KK, Basu M, Chakraborty N (2008) Particle swarm optimization technique based short-term hydrothermal scheduling. Appl Soft Comput 8:1392–1399
33. Mandal KK, Chakraborty N (2013) Parameter study of differential evolution based optimal scheduling Of hydrothermal systems. J Hydro-Environ Res 7:72–80
34. Basu M (2014) Improved differential evolution for short-term hydrothermal scheduling. Electr Power Energy Syst 58:91–100
35. Sivasubramani S, Swarup KS (2011) Hybrid DE–SQP algorithm for non-convex short-term hydrothermal scheduling problem. Energy Convers Manag 52:757–761
36. Basu, M (2014) Artificial bee colony optimization for short-term hydrothermal scheduling. J Inst Eng (India): Ser B 95(4):319–328
37. Gouthamkumar N, Sharma V, Naresh R (2015) Hybridized gravitational search algorithm for short-term hydrothermal scheduling. IETE J Res. https://doi.org/10.1080/03772063.2015.1083904:468-478
38. Dubey HM, Panigrahi BK, Pandit M (2015) Improved flower pollination algorithm for short term hydrothermal scheduling. Lect Notes Comput Sci 8947:721–737. https://doi.org/10.1007/978-3-319-20294-5_62
39. Orero SO, Irving MR (1998) A genetic algorithm modeling framework and solution technique for short-term optimal hydrothermal scheduling. IEEE Trans Power Syst 13(2):501–518

40. Lakshminarasimman L, Subramanian S (2006) Short-term scheduling of hydrothermal power system with cascaded reservoirs by using modified differential evolution. In: IEE proceedings—generation, transmission and distribution, vol 153(6), pp 693–700. https://doi.org/10.1049/ip-gtd:20050407
41. Lakshminarasimman L, Subramanian S (2008) A modified hybrid differential evolution for short-term scheduling of hydrothermal power systems with cascaded reservoirs. Energy Convers Manag 49:2513–2521
42. Nguyen TT, Vo DN, Dao TT (2014) Cuckoo search algorithm using different distributions for short-term hydrothermal scheduling with cascaded hydropower plants. https://doi.org/10.1109/TENCON.2014.7022454
43. Hota PK, Barisal AK, Chakrabarti R (2009) An improved PSO technique for short-term optimal hydrothermal scheduling. Electr Power Syst Res 79:1047–1053
44. Chang W (2010) Optimal scheduling of hydrothermal system based on improved particle swarm optimization. https://doi.org/10.1109/APPEEC.2010.5448307
45. Sinha N, Chakrabarti R, Chattopadhyay PK (2003) Fast evolutionary programming techniques for short-term hydrothermal scheduling. IEEE Trans Power Syst 18(1):214–220
46. Bhattacharya A, Bhattacharya A, Datta S, Basu M (2013) Gravitational search algorithm optimization for short-term hydrothermal scheduling. https://doi.org/10.1109/ICETEEEM.2012.6494479
47. Rao RV, Savsani VJ, Vakharia DP (2011) Teaching-learning-based optimization: a novel method for constrained mechanical design optimization problems. Comput Aided Des 43(3):303–315

The manufacturer's authorised representative in the EU is Springer
Nature Customer Service Centre GmbH, Europaplatz 3, 69115 Heidelberg,
Germany. If you have any concerns regarding our products, please
contact ProductSafety@springernature.com

Printed and bound by CPI Group (UK) Ltd, Croydon, CR0 4YY
30/09/2026
02225574-0004